CROSSING THAT BRIDGE

A Critical Look at the PEI Fixed Link

EDITED BY LORRAINE BEGLEY

To all Maritimers who spoke or acted against the process which denied us the choice of transportation options and the right to legitimate inquiry and informed debate.

COVER PHOTOGRAPH:
Wayne Barrett — Marine Atlantic

PRINTED AND BOUND IN CANADA BY:
Les Éditions Marquis Ltée

PUBLISHED BY:
Ragweed Press
P.O. Box 2023
Charlottetown, P.E.I.
Canada C1A 7N7

DISTRIBUTED IN CANADA BY:
General Publishing

Thanks to the Cooper Institute and the Canadian Labour Congress for kind permission to reprint a revised version of "The Economic Impact of the Withdrawal of Marine Atlantic Ferry Service from PEI." Thanks also to the *Eastern Graphic* for generous permission to reprint selected editorials.

CANADIAN CATALOGUING IN PUBLICATION DATA:
Main entry under title:

Crossing that bridge

Includes index.
ISBN 0-921556-39-X

1. Bridges — Prince Edward Island — Planning.
2. Transportation — Social aspects — Prince Edward Island. I. Begley, Lorraine, 1951-

HE377.C32P7 1993 388.1'32'09717 C93-098544-3

Contents

Island Society 112

The Government, the Process and the People 146

"... decisions we as Islanders make on this momentous and grave matter must be the right decisions. We owe this to ourselves, to our children and to all future citizens of this province. Mr. Speaker, there is no room for error, no opportunity for second guesses and little latitude for corrective action if our planning is inadequate and unable to meet the test of time. Simply put, Mr. Speaker, we must know what we are doing."

Former Premier Joseph Ghiz, in a reply to a speech from the throne in the PEI legislature, 20 March 1987.

Introduction

This book is not about the "Island Way of Life."

The notion of such a way of life — distinct from that of the mainland, maintaining an emphasis on the integrity of communities and families and the interdependence of relationships — has been central to oppositional politics on Prince Edward Island in recent years, including the politics of opposition to the proposed mega-bridge across the Northumberland Strait. The promise of this idea has, however, largely been exhausted.

About a quarter century ago, as part of the Development Plan which put an end to an earlier fixed link causeway-construction project, a decision was made by the dominant elite that Prince Edward Island should abandon emphasis on its traditional resource-based economy of farming and fishing and concentrate instead on the service industry — on tourism.

Part of the process of "selling" Prince Edward Island to a largely urban market in central Canada and the northeastern United States was to "kidnap" the structure of ideas and cultural forms by which Prince Edward Islanders think of themselves. In the hands of the elite, the people's shared vision of themselves was transformed into offensive characterizations and alien representations.

Prince Edward Island was touted as a paradise of traditional communities and values, and all the vacuous cultural dichotomies that typically give shape to this sort of cultural imperialism were imposed on it: Mississauga and Oshawa are "modern"; Minemegash and O'Leary are "traditional." New Yorkers and New Jerseyans travel in the lane of the fast and frantic; those who live in New Haven and New Dominion are permanently nestled into the slow-paced timelessness of the shoulder. Scarborough and St. Catherines are wrapped up in the isolating world of the self-propelled individual; Summerside and St. Eleanors summon the traveller to cocoon-like communities, to a place that is still a home. At the whim of the socially dominant, words such as "home" and "community" became, in Prince Edward Island, the words of alienation. As the currency of the Department of Tourism, they have been drained of their meaning and then refilled, bloated, discoloured.

Articulators of the "Island Way of Life," in trying to build an oppositional movement using the semantics of the Department of Tourism, have become stuck fast in the quicksand of those semantics. As a result, proponents of the "Island Way of Life" were seen as "backward," while pro-bridge boosters were allowed to cast themselves as the voice of progress, improvement, modernization and the future. The proponents of the bridge managed to marginalize an opposition expressing itself in the language of community, home and the "Island Way of Life," as nostalgic echoes of the traditionalism and quaintness to which, paradoxically, their fixed crossing will improve market access.

And so this book is not about communities, homes, or a distinctive Island way of life. This book is about a bridge. Or rather, the opposition to a bridge. Most Canadians know something about the Oldman River Dam, a project which proposes to flood 5,800 acres of southwest Alberta at a cost of $355 million. Most know that it is opposed by environmentalists and the Peigan Indians, part of whose land will be flooded, because important wildlife habitats, archaeological sites, historic sites, graves and other sacred places of the Peigan Nation will be destroyed.

Many Canadians know something about the Rafferty-Alameda project which proposes to divert fresh water from Saskatchewan to the United States for the next 100 years at a cost of $150 million. And they have heard that it is opposed by farmers, ranchers, aboriginal peoples, church and community peace and justice activists.

Almost all Canadians, many other Maritimers and some Islanders are unaware that the proposed bridge between New Brunswick and Prince Edward Island is opposed by a great number of people on the Island because of large-scale social, economic and environmental damage which would occur if the 13 kilometre bridge across the Northumberland Strait, which separates New Brunswick from Prince Edward Island, is allowed to proceed. *Crossing That Bridge: A Critical Look at the PEI Fixed Link* is about that opposition.

Opposition to this project comes from many corners. In 1987, the first organized opposition formed to respond to a plebiscite. This vote was called by the provincial government in the face of a growing and vocal alarm at the blinkered fast-track method of proceeding which has come to characterize the federal government's attitude toward this mega-project. As more information on the proposal and its impact became available, and as the political manoeuverings of the federal government became clearer, others moved to question the necessity of the mega-project. Scientists such as Irené Novaczek, Mike Dadswell, Owen Hertzman and Erik Banke point to the monstrous environmental consequences of building a bridge across the Strait — consequences which speak of global inter-relatedness and which will mean both localized and global environmental change from seemingly confined actions.

Other opponents studied the economic and financial details of a mega-project which is supposed to be a privately financed, economic bonanza for Prince Edward Islanders. Don Deacon, Peter Townley, Joseph O'Grady and a report by the Cooper Institute in Charlottetown scratch past the shallow veneer of economic and financial viability, exposing the shoddy under-the-surface structure of this project.

The issue of privatization of transportation infrastructure has not been widely discussed, yet it figures prominently among the concerns about the fixed link which will have resonance throughout the country. The mythology of privatization, says Sharon Fraser, is a child of the '80s — the decade of greed. In reality, privatization of the fixed link means that "We allow a small self-interested group to purloin what has always been ours ... and then make us pay for the privelege of using it."

Reg Phelan and Kevin Arsenault share common ground in discussing Prince Edward Island's traditional concern with land use and ownership, and uncontrolled tourism growth — two dangers resulting from the sticky combination of a provincial government that is politically soft on both issues and the construction of a privately-owned fixed link that is meant to generate profits.

The likelihood that the proposed bridge will affect many of Prince Edward Island's 130,000 residents in some way seems to have been lost on the local media who, according to Martin Dorrell's report card on them, should receive marks that range from C to F. The national media ranks at the bottom — their reporting was shoddy, even when they did manage to show up.

The machinations of the federal and provincial governments and the development consortium comprise a saga of deceit, dishonesty and undemocratic action which Jim Brown details in his study of the process by which a fixed link has been chosen. Daniel Schulman and Tom Kierans each outlines thoughtful proposals on avoiding a repeat of these manipulations and, at the same time, provide a thorough, accurate and fair assessment of this and future environmental questions.

I am indebted to many people for their advice and encouragement during the preparation of *Crossing That Bridge*. I would like to thank Louise Fleming, Lynn Henry and the rest of the staff at Ragweed Press for giving me the opportunity to compile into a permanent record the words and worries of Maritimers who have, after considerable research and thought, spoken and written their opposition to the mega-bridge project. The response to my requests for detailed essays in a short time frame has been heartening, and the dedication writers have exhibited to this project is beyond description.

I would also like to thank Jim MacNeill and Jack McAndrew for permission to reprint excerpts from columns and articles which originally appeared in the *Eastern Graphic*.

For advice, ideas and generosity in discussing the fixed link with me — in one case, literally until the cows came home — I owe much to Maureen Larkin, Ruth Freeman, Captain Donald Graham, Douglas MacFarlane, Harry Baglole, Catherine Edwards, Carol Livingstone, Maurice and Jean Lodge, Anne Lie-Neilsen and my neighbours and (still) friends, Annie and Ewen MacPhail.

The staff at the reference desk at both the Robertson Library at UPEI and at the Confederation Centre Public Library have been invariably helpful to me in confirming details and assisting in my research.

I would like to thank my friends and editors at *New Maritimes* for giving me the courage to take on this book project. Gary Burrill, Ken Clare and Scott Milsom, with whom I have worked closely over the years, will know of the anguish I experienced in compiling this book.

I would like to thank my family — my children Josie, Katie Colleen and Jean for their patience and understanding when, though "at home," I was unavailable to them, and for their questions, which forced me to clarify my thoughts and feelings about the fixed link. The patience, encouragement and critical comments of my husband Richard Baker have made my job immeasurably easier. I would also like to thank him for his understanding when my writing made me a not-very-fun-person to be around.

My deepest gratitude is to the writers who gave so freely of their time, research and expertise: thank you.

L.B.
Argyle Shore
April 1993

History

"You pay a price for progress and economic change. And I believe the best interests of the Island are served by the most efficient, modern communications with the mainland in every respect; transport, telecommunications and so on. And there's bound to be some changes as a result of this but I believe they'll be positive. They may change the way of life to some extent but governments can compensate for this."

Elmer MacKay, Minister of Public Works,
in conversation with CBC Radio, 3 December 1992.

There have been times over the past years when Islanders, in their individual disgust with the deafening drone of commentary on the fixed link issue, have switched off their radios and televisions, refused to buy newspapers if the words "fixed link" appeared anywhere on the front page and covered their ears to coffee-shop conversation if those dreaded words came up.

These protests go to the heart of the fixed link process, to a core of resentment about being isolated from the project, about having no voice, no input, about the people's opinion not mattering to the outcome of the project.

The decisions about our transportation system have been made elsewhere, presumably by people who know better than Prince Edward Islanders what best suits us. The four options available for crossing the Strait — improved ferry service, road bridge, road tunnel and rail tunnel — have been manipulated and narrowed, (but never, never, compared) to the bridge-only option. Although improved ferry service might be the most apt crossing method for Prince Edward Island, it rules out something dear to Public Works Canada (PWC) and corporate Canada's heart — a mega-project, the bigger the better. This manipulation didn't go unnoticed.

There has been no shortage of words about a fixed link — words broadcast and printed, repeated and refuted, shouted and stifled. Despite this outcry, there continue to be raw ragged holes in the fabric of our knowledge. One of the most obvious, and one which remains despite efforts by a number of menders and patchers to cover it over, is the question of a more literal hole — a tunnel.

The tunnel issue has become an angry spectre which haunts every corner of this bridge project. Nowhere can Public Works Canada hide from the cursed, unceasing chant: "Wwwwwwhat hhhhhhhappened to the tttttunnel cccccconcept?" The weekly press rhetorically asked the question in 1988. The federally appointed Environmental Assessment Panel (EAP) formally asked the question of PWC in 1989. Speaker after speaker at the EAP hearings in 1990 asked the question. Fishers asked it, truckers asked it, farmers asked it, homemakers, poets, teachers, letter-carriers — they all asked, "What happened to the tunnel concept?" Public Works Canada still refused an explanation. In March 1993, in the Federal Court of Canada, PWC was asked again, "What happened to the tunnel concept?" Even the national press has reported that no plausible answer has yet been provided.

The Fix Is In: An Overview of the Fixed Link Fiasco

Lorraine Begley

"Fixed link" have been such convenient words for Public Works Canada (PWC). The words belie the long-held intention of this government department to build a bridge and nothing but a bridge across the 13 kilometre width of Northumberland Strait. In less politically charged times, the words would have meant the variety of crossing options, with the exception of the ferry service — a bridge, a highway tunnel or a rail tunnel.

Even today, with the "bridge" project successfully manoeuvred into the winner's circle, the words "fixed link," retain their aura of having offered a choice to Islanders, and still conveniently mask the growing ambiguities surrounding the "bridge" construction project.

Somewhere along the line, the stated rational for building a fixed link — to save money and improve transportation between Prince Edward Island and the mainland — was drowned in the frenzied hyperbole of a business sector hungry for a sniff of the short term economic boost the mega-bridge will undoubtedly give the regional economy. In the face of growing proof against the wisdom of constructing a bridge across the Northumberland Strait, these short term economic benefits now represent the largest rationale for the project's continued presence on the federal and provincial political agendas.

Since the idea of this particular fixed link (this is the fourth time a proposal to build such a link has entered PEI's history) first emerged in the early spring of 1985, Prince Edward Islanders have worked hard, both as individuals and as members of different organizations, to examine the issue: The many reports on the subject were followed by various responses. Public hearings, open houses and information sessions were followed by press releases, interviews and sound bites, each with conflicting interpretations. Opponents of the project are becoming exhausted by the lengthy process, and its proponents ... well, they just enjoy the steady march the project is making toward completion.

Opponents are angered by the way the federal government, through Public Works Canada, appears to be ramming through the plan for a bridge. They talk about problems with ice build-up, monstrous damage to the environment and tears in the social fabric and community life of the Island. Proponents forecast a new Eden if only we could get our potatoes over there, and get "them" (tourists, visitors from the mainland) over here, more easily and quickly.

Much of the information that has been amassed on the "fixed link" resides in the Northumberland Strait Crossing Office on Queen Street in Charlottetown. This office is PWC's public relations/information bureau on PEI. There, copies of a long list of studies, reports and documents published by PWC are available — reports which are laden with labels like "High Resolution SAR Data Acquisition and Interpretation for Northumberland Strait" and "Dynamic Behaviour of Northumberland Till."

There, also, is the "Report of the Environmental Assessment Panel," a document which resulted from over a year of study and public hearings into PWC's generic bridge proposal. The August 1990 report is quite explicit and clear-cut about the proposed bridge: On two occasions the report said, in bold print, "The Panel recommends, therefore, that the project not proceed." Consideration, it continued, should be given to alternative ways of improving transportation across the Strait, such as better ferry service or, possibly, a tunnel. The Environmental Assessment Panel (EAP) believed a bridge would make too much of a mess of the environment in an area some have called the only bright spot in an otherwise dismal Atlantic fishery — the shell fishery in the Northumberland Strait.

Within a few months of its publication, however, the Report's conclusions and recommendations had been all but forgotten by PWC, and the bridge project was again being rushed towards a precipice from which, it seemed, there could be no turning back — namely, the selection of the winning consortium, to be followed by the awarding of a construction contract.

With environmental researchers saying "no" to a bridge and "yes" to alternatives, and PWC saying "yes" to a bridge and "no" to everything else, the project — which different estimates say will cost anywhere between $800 million and something close to twice that amount — seems a bit bizarre.

Published reports, most commissioned by PWC, evidenced glaring problems with information and interpretation and even consistency. For example, two studies titled "Economic Feasibility Assessment" and "Financial Analysis," both contain information based on the first 35 years of the project. Figures in the two reports that reflect the operational costs of keeping the Marine Atlantic ferries running between Borden and Cape Tormentine differ by $5 million. By exaggerating Marine Atlantic's operating costs and underestimating its steadily increasing income, the two studies manage to inflate the cost of keeping the ferry service to between $15 and $21 million above Marine Atlantic's own estimates.

Estimates of the operating costs for the proposed bridge are suspiciously low too. If PWC is right in its estimate that the bridge will provide 70 permanent full-time jobs, the projections in the "Economic Feasibility Assessment" show that annual salaries approximately $6,000 below what Marine Atlantic workers now receive will swallow the entire operating budget. If, on the other hand, the projections in the "Financial Analysis" are correct, salaries of $14,500 per year will consume all of its estimate of operating costs for the bridge. Absent from both the studies' calculations are any expenditures associated with bridge maintenance — costs that some sources say will be astronomical.

To jaundice the issue still further, the operating costs for the bridge were placed side-by-side in both reports with the total costs of capital, maintenance and operation of the ferry service — a sleight of eye that encourages the proverbial apples-and-oranges comparison.

On the subject of proverbs, PWC has a new angle on the "Phantom Ship" that is said to sail the Northumberland Strait and has been immortalized in song by Island musician Lennie Gallant. Included in the calculations contained in the "Economic Feasibility Assessment" is the cost of an additional ferry — not just the replacement of an existing one. Yet Transport Canada doesn't deem this spectral vessel necessary in its long-term estimates, nor does Marine Atlantic. Fiander-Good Associates, the consulting firm which authored the report, explains that the figures for its financial analysis were provided by Transport Canada "with some input from Marine Atlantic and PWC." If neither Marine Atlantic nor Transport Canada feel that an additional vessel will be required in the next 35 years, this leaves only PWC's "input" to account for the statement in the "Economic Feasibility Assessment" that "this vessel is considered necessary by Fiander-Good Associates Ltd. to meet capacity requirements near the end of the study period." This ghostly vessel substantially increases an annual "subsidy" that Ottawa will be doling out to the developers, a yearly payment of $42 million for a period of 35 years to Strait Crossing Inc. (SCI), the consortium picked to construct and operate a bridge.

There is considerable confusion about just how the $42 million figure was arrived at. There are, apparently, two ways of calculating this "subsidy": the obvious way, and PWC's way. The way most people would understand the word "subsidy" — and this is also the way Marine Atlantic's annual subsidy from Ottawa is figured — is as the difference between the annual operating costs, plus capital expenses, minus revenues. Marine Atlantic has estimated that this will average about $31 million annually over the next 35 years and covers capital expenses such as new boats. The annual operating subsidy it receives from Ottawa has been declining in recent years — this has been offset by increasing revenues. Marine Atlantic has taken exception to the federal government's plan to offer a $42 million annual subsidy to the bridge's developer: "We *cannot*, and *do not* accept that [$42 million] estimate," says Marine Atlantic Vice-President Murray Ryder, adding that it "is totally at odds with what has actually happened in the past."

For PWC and the federal government, the "subsidy" is thought of as something quite different. According to PWC's "Financial Analysis," prepared by Woods-Gordon Management Consultants, the amount of the "subsidy" that will go to the developer is not directly tied to Marine Atlantic's projected subsidy needs. It is, rather, a compilation of dollars from a number of questionable sources. One of the many sets of calculations used to arrive at the $42 million annual figure is based on overestimates of Marine Atlantic's operating costs and underestimates of its steadily increasing revenue stream. Another component of the proposed "subsidy" is labelled as "Subsidy Available for Bridge," that is, extra money the government is willing to simply transfer to the developers. The federal government is also going to throw into the developers' pot those revenues "generated by the sale of Borden ferry assets." And, the revenues from the sale of one of the Wood Islands ferries are also factored into the calculations that cooked up the $42 million figure, as are the costs that would have been incurred, without a fixed link, to pay for vessel replacement and for making repairs and improvements to docks and harbours.

When a bridge is built, what will become of the Wood Islands-Caribou service that has long provided a major link between eastern Nova Scotia and Prince Edward Island? In September, 1985, the federal government's Task Force on Program Review was chaired by Erik Nielsen, the same person who, four years later, presided over the abandonment of rail service on PEI. One of the Task Force's recommendations was that the federal government commence "negotiations with Prince Edward Island ... to end federal subsidies for the 'alternates' to constitutional services" — that is, the Wood Islands ferry service.

The construction of a fixed link will serve this purpose nicely. The calculations in the "Financial Analysis" show that capital expenditures for the Wood Islands service will be postponed, and then eliminated. The Wood Islands service will be downgraded. Initially, some traffic will still choose the Wood Islands ferry, but later, PWC's projections show that there will be a decline in federal subsidies. This will be offset by a combination of increased fares and reduced service. And the money saved will go directly to the developers as part of the federal "subsidy."

Ottawa can eliminate the Wood Islands service in one of two ways. It could follow the strategy it used so effectively to achieve rail abandonment on the Island, termed "demarketing": The Wood Islands service would simply be "demarketed" to the point where it became too expensive and inconvenient to use. Then, invoking its "use-it-or-lose-it" stratagem, the federal government would shut it down.

The second way Ottawa might kill the Wood Islands service is broadly hinted at in a section of PWC's "Financial Analysis" entitled "Subsidy Available to Bridge by Closure of Wood Islands Ferries": Shut the service down directly and give the money directly to the developers.

• • •

As odd as PWC's ideas about "subsidies" and "private" financing are, its actions show it to be covert and incompetent about the economic feasibility of the bridge project.

And it isn't eager to talk about it: Until January 1992, PWC, claiming they were "confidential," refused requests made under the Access to Information Act for documents on how the Department of Finance feels about the project. At that time the federal government declared its support for the project, claiming it was the kind of private initiative mega-project the federal government wanted to encourage. One year later, in January 1993, and only because legal proceedings forced it to, PWC released a document by Gardner Pinfold Consulting Economists which confirmed criticisms of the government's economic feasibility study and of PWC's interpretation of it.

This had-be-released report says that the economic feasibility study "does not establish the economic viability of the fixed crossing. If economic viability is an important criterion for proceeding with the project, then it would be advisable to conduct a more rigorous analysis."

And PWC, I discovered, is touchy about another subject as well — tunnels. Requests for documents relating to proposals to build a tunnel beneath the Strait have been refused to the point that PWC denies that any such documentation exists. However, a number of such documents came into my hands from another source. Through these, I learned that PWC has an aversion to both road and rail tunnels. Jim Feltham is one of three PWC staff who reviewed, and then denied, my request for documents on the tunnel-crossing option. He is also Project Manager for the Northumberland Strait Crossing Project.

A number of summers ago, Feltham wrote several letters to Jim Clarke, Executive Secretary of the Federal Environmental Assessment Review Office (FEARO). The first of these outlined the reasons why a road tunnel was rejected as a feasible option by PWC. A second letter listed six reasons for rejecting the idea of a rail tunnel.

In 1987, PWC made it known that it would not consider a rail tunnel as a crossing option. This decision was made on the basis of a single unsolicited proposal it had received in 1985. That proposition was put forward by a consortium called Omni System Group and involved a single-track rail tunnel with a passing track at mid-point under the Strait. Trains would leave each side every 15 minutes and make the crossing in 13 minutes, both winter and summer. Omni System asserted that there would be enough capacity to ensure that vehicles would not be left behind to wait for the next train, that delays caused by weather would be minimal, that its proposal imposed "no additional impacts on the environment, and eliminates many environmental impediments to safe and efficient crossings," and that capital costs would be "substantially below all other fixed crossing proposals." And, PWC's own "Economic Feasibility Assessment" showed that annual operating costs for the proposal, in 1991 dollars, would be less than the amount of operating revenue Marine Atlantic predicts will be generated by its ferry service. In other words, the operation of the rail tunnel would be virtually self-sufficient.

These points were not directly challenged by PWC. Instead, the rail tunnel proposal was rejected for the six reasons given in Feltham's second letter to Jim Clarke.

An examination of these purported reasons is revealing. Two of them, Feltham wrote, concerned certain calculations PWC made called "net present value of cost" and "net present value of benefits." These calculations sprang, at least in part, from the exaggerated figures used to pad the operating costs of the ferry service and from glib assumptions and questionable data. For example, it was assumed, for no apparent reason, that the combined waiting and crossing time for a rail tunnel would be three times that of a bridge.

They were also based on a controversial analysis of what is "benefit" and what is "cost." Even the authors, it seems, of the PWC's "Economic Feasibility Assessment" had difficulty trying to justify these questionable calculations — others, such as Gardner Pinfold Consulting Economists, who criticized this study, don't try. They say simply "The approach used by Fiander-Good to quantify benefits for induced traffic is not correct."

Another of the six reasons given for rejecting the rail tunnel project was that "it offered little improvement over the existing ferry service." This is untrue. Even the biased calculations of the "Economic Feasibility Assessment" said "a rail tunnel will result in a 55-minute time saving" compared to its estimate of a 100-minute average crossing/waiting time for the Borden-Cape Tormentine ferry service.

Feltham's letter also expressed concern about the project's carrying capacity and feared that it would be "exceeded shortly into the 35-year developer operating period." Omni System's proposal claimed that capacity would be "sufficient for long-term growth forecast." This question still seems unresolved.

One, at least, of Feltham's reasons for rejecting the Omni System proposal was more to the point. He wrote, "It was projected in 1987 that the PEI rail service would be discontinued within the next five years." Some might ask: So what? The proposal, after all, was for a rail tunnel that would transport cars and trucks from one side of the Strait to the other. It was never intended to make it easier to haul freight from Toronto to Tignish.

But Jim Feltham's mention of the Island's now-defunct rail service underlines the fact that the Omni System proposal was a victim of, more than anything else, bad timing. Around the time of the proposal's submission in 1985, the federal government was promoting a document called *Freedom to Move,* which outlined the new Tory philosophy for transportation: in a word, deregulation. *Freedom to Move* outlined the imposition of a conservative, private-enterprise, corporate agenda for Canadian transportation reform. One of the then-unstated items on that agenda was the phasing out of Island rail service. Although a joint federal-provincial report released in early 1988 described the cost to Island industry of rail abandonment as "considerable" and warned further that "considerable industrial development impacts due to increased transportation costs" might be expected on the Island if the trains ceased to roll, the Tory government persisted with its plan. Over the years, train freight service was eroded — or, more euphemistically, "demarketed" — until it was finally terminated at the end of 1989.

Omni System's proposal had the ill luck to come along just as national transportation policy was moving strongly in the direction of privatization and deregulation, and it held out the hope of a shot in the arm for the railroad on PEI. Although they intended to provide electrically powered, rather than diesel-powered, rail transportation across the Strait, the proposal's authors made the political mistake of stating that, as a bonus, "rail freight can be drawn by electric power." This would open the door for increased rail service on PEI. With a rail tunnel, a "continuous crossing" of the Strait would be possible for the first time in history. This opened up numerous possibilities that a government set on deregulation would not be keen on, including increased rail freight and perhaps even a passenger service that would connect to the main line at Moncton. And, although the Omni System proposal would be operated as a private enterprise for a period of time, ownership would eventually revert to the federal government and it would be back in the rail transportation business.

To Tory privatizers and deregulators, this would be a nightmare-come-true. They knew in 1987 that their plans for rail abandonment would be resisted on the Island, and they had no intention of offering ammunition to the opposition: The rail tunnel was dropped as a fixed link option. (It is ironic that the dismantling of the abandoned rail bed and track from Moncton to Cape Tormentine has been halted in the event that the line might be needed to transport materials to the site of bridge construction.)

In her book, *Parcel of Rogues: How Free Trade is Failing Canada*, Maude Barlow writes:

> No other nation in history has ever deliberately destroyed its railway system except to forestall invasion during wartime. The government claims that railways are a luxury we can no longer afford. But their cost is small compared to those of the alternatives: greatly expanded highways, upgrading airports to handle increased traffic, the layoffs of thousands of employees, and the isolation and impoverishment of whole communities. Who is counting the environmental impact of shutting down a clean mode of transportation when all the world is seeking to create one? ... Will the real Canadian position on global warming please stand up?

Unfortunately, neither PWC nor the Mulroney government were compelled to answer questions like Maude Barlow's. And so, the most economical, environmentally appropriate, far-sighted and flexible approach to a fixed link was eliminated, not out of the federal government's desire to make the best choice for PEI, but simply because it conflicted with the Tory political agenda.

As yet, mention has been made of only five of the six reasons outlined in Feltham's letter for dropping the rail tunnel option. The last one is that a rail tunnel would offer only "intermittent service." A bridge, in contrast, boosters of that concept claim, would offer "continuous service." After all, this logic goes, even if the rail service would get you across the Strait in just 13 minutes, the train will only leave every 15 minutes, so you might have to wait up to that much longer to get "on the road again." With a bridge, it's just a matter of driving on and driving off, and so it's that much quicker. Or so the theory goes ...

• • •

Just imagine: It's mid-June, 1999, the bridge is completed and you're heading to the mainland. The summer sun is warm and there's a light breeze off the water. As you head out onto the bridge you roll down the window and let the wind cool you. The scene is blissful as you survey the red cliffs of your home province from your privileged vantage point. Your thoughts wander back to the days of your grandparents and great-grandparents, when, until a ferry steamer was finally introduced during World War I, winter crossings were hazarded only by small iceboats. "What a difference," you think, gazing again at the marvelous panorama. Then, all too soon, you descend into New Brunswick and make your way west. The crossing was, well, like a dream.

In fact, such a crossing will be little more than a dream most of the time.

As part of its contribution to the fixed link literature, PWC has produced a publication called *Strait Facts* which includes a cute illustration showing a cute small car and a cute truck crossing an equally cute bridge that spans slightly rippled water. An accompanying cute graph shows the "actual crossing time" to be 13 minutes. But just how realistic is "cute?"

The "slightly rippled water" in the illustration is more important than it might seem, because in the building and operation of any bridge, wind is a key factor. So, an understanding of wind dynamics is essential to any understanding of how a bridge is operated. One central aspect of these dynamics is that the higher you get from ground or sea level, the higher wind speeds tend to be. Another is that wind speeds usually increase over open water.

The planned bridge will have two 3.75m wide lanes plus 1.75m wide shoulders, and will measure 27 kilometers, including approaches. The over-the-water section will be 13 kilometres long, 64m (210 feet) high at the central span, and 40m (135 feet) high on the adjoining side spans.

Late in the fall of 1988, Jim Feltham and three other PWC officials made a "technical visit" to the 8 kilometre Mackinac Bridge in Michigan. In a report made following their visit, they said, "The 31-year-old Mackinac Bridge offers striking ice climate, wind and socio-economic parallels to the proposed Northumberland Strait crossing ..."

In fact, a significant leap of fancy is required to see "striking parallels" between the Mackinac Bridge and the one planned for the Northumberland Strait. Especially fanciful are parallels between the socio-economic situation of PEI and that of northern Michigan, between ice and wind in a freshwater lake and the Northumberland Strait's often massive, tide-tossed pack ice and wind that, to quote one wit, "almost blows the hair off a dog." As for parallels — at least the centre spans are strikingly parallel in height — well, almost.

A few days before Christmas, 1988, just about a month after Jim Feltham returned from the Michigan trip, he received a PWC report entitled "Investigations into the Effect of High Winds on Vehicular Traffic." That, together with another report handed to him the previous year entitled "Assessment of Winds, Waves, Tides and Currents," made Jim Feltham aware that wind conditions of the Northumberland Strait bear no resemblance to those at the Mackinac Bridge. Yet, three months later, when their report on the Mackinac Bridge visit was published, this knowledge was ignored by the authors who penned it.

Here are some realistic comparisons. Find parallels, if you can.

In the 5 years that wind speeds have been monitored by an anemometer mounted 3m above road level on the Mackinac Bridge — 63m above water level — the maximum wind speed recorded was 115 kph. Winds of that speed can be expected in the Northumberland Strait at only 10m above water every 5 years. (Environment Canada considers winds of 118 kph to be of hurricane force.) PWC calculations of wind speeds neglect to show the frequency of hurricane-force winds *above* 10m: They can be expected so frequently that PWC's wind-speed studies don't include them.

In 1975, the lake boat *Edmund Fitzgerald*, remembered in song by Gordon Lightfoot, sank in Lake Michigan, taking the lives of its entire crew. That November night, winds blew to 130 kph, the most severe ever recorded in the 34-year history of the Mackinac Bridge. PWC data shows that *Edmund Fitzgerald*-calibre winds can be expected in the Strait every 5 years at a level of only 40m. And, wind *gusts* to such speeds can be expected so often at only 10m above water that PWC hasn't even bothered to estimate their frequency. Despite these PWC calculations, one of its spokespersons assures me that winds at the Mackinac Bridge were more severe than those in the Northumberland Strait.

Located near the back of the "Investigations into the Effect of High Winds on Vehicular Traffic" study is a chart of wind speeds 50m above water — that is, a full 14m, or 46 feet, below the height of the planned bridge's central span. But the chart is nonetheless instructive: Even at this low level, 50 kph winds can be expected 21% of the time, while 40 kph-plus winds will blow 42% of the time.

"High winds" — defined in this study as those above 40 kph — will necessitate "special management control measures" because "the overturning and handling stability of vehicles in high winds are [sic] enhanced by a reduction in vehicle speed." Short of closing the bridge, these measures will include convoying vehicles and reducing the speed of traffic. Thus, traffic speed will be reduced to 40 kph when winds blow between 40 and 50 kph. Crossing the over-the-water part of the bridge will now take 22 minutes. At wind speeds of between 50 and 70 kph, "empty semi-trailers, campers and light vans [will be] escorted in convoy." In 70 to 100 kph winds, an escorted crossing of "all vehicles" at 30 kph will take 26 minutes. ("All vehicles," that is, except "empty semi-trailers, campers and light trucks," which will have to cool their tires in the parking lot until winds subside.) At 100 kph, consideration will be given to closing the bridge or, alternately, to escorting "permitted" vehicles at 20 kph — making for a harrowing 39-minute crossing, plus queuing, convoy assembly time, and so on.

These safety measures mean that convoys will be necessary whenever wind speeds exceed 50 kph, which is, even 14m below the central span, more than 20% of the time. Traffic travelling behind a convoy will have to travel at the reduced convoy speed as well. It is possible that only one convoy will be able to use the bridge at a time: The one travelling in the opposite direction will have to wait for the arrival of the first before beginning its snail's-pace journey to the farther shore. More likely than this, however, all traffic will be slowed due to the frequent necessity to convoy, because, at bridge height, winds will be in the 50 kph range more than 20% of the time — much of this in the October to May period when wind often brings with it precipitation in its myriad forms.

Future patrons of the bridge-to-be will be interested to know that, even under normal operating conditions, trucks on the Mackinac Bridge must limit their speed to 30 kph, that passing is prohibited and that trucks must travel at least 150m (500 feet) apart. According to PWC, the Mackinac Bridge has an excellent safety record. This is due, in no small part, to its strict safety regulations, precautions that federal specialists should ignore only at their peril. Under the best of operating conditions, traffic travelling behind any of the 160,000 trucks that cross the Strait every year will have made the trek at the same speed as the truck. That's 30 kph, making for a 26-minute crossing of the over-the-water section of the bridge, and 54 minutes for the 27 kilometre length of the bridge plus approaches. Add a touch of snow in the air and winds of any speed, and crossing the bridge will require sober consideration.

What PWC's *Strait Facts* illustration actually shows is a pie-in-the-sky vision of what a bridge crossing will be like. But it is a vision that PWC is determined to preserve in the public mind. The illusion of a "continuous crossing," with a delightfully sunny drive and a fabulous view is, in reality, a severely handicapped transportation link and, at best, "intermittent service."

• • •

It's obvious that PWC rejected the idea of a rail tunnel for political reasons. But what of a road tunnel? Did PWC ever even consider it as an option?

Project Manager Jim Feltham said in a June, 1989 letter to Jim Clarke, Executive Secretary to the FEARO, that: "The response to Public Works Canada Stage 1 Call for expressions of interest had resulted in seven developer consortia each of which carried demonstrated capability in the construction of both bridges and tunnels."

Public Works Canada recognized that the "acid" test for a tunnel would be the response of the private sector to the Proposal Call. Of the seven developer consortia invited to submit proposals, one developer submitted a tunnel proposal, namely PEI Crossing Ventures Ltd. (Lavalin). They also submitted a bridge proposal.

W.A. Stephenson Construction (Western), Limited is a partner of Strait Crossing Inc., the consortium selected by PWC to build the mega-bridge across the Northumberland Strait. W.A. Stephenson Construction publishes promotional material with full-colour pictures and text, advertising its expertise in tunnel construction. It has this to say about its tunnelling abilities:

> Experience gained in the construction of more than 120 kilometres of underground tunnels has provided us with a solid background in this particular field of construction. Our technical know-how includes ... bored and blasted tunnelling techniques. In many instances, advanced new construction techniques, which may be regarded as pioneer, have been used, as well as conventional construction methods.

With experience like this why would W. A. Stephenson Construction submit a proposal for a bridge, with all its attendant environmental, wind and ice problems, rather than a tunnel proposal?

Tom Kierans, a tunnelling expert retained by Environment Canada to advise Environmental Assessment Panel (EAP) members on technical matters, says simply that PWC's Proposal Call to developers was biased — it favoured a bridge over a road tunnel. The importance of bias in PWC's actions has never been adequately addressed. Its methods were twofold: they involved both acts of commission and acts of omission.

Under the heading "Acts of Commission" falls PWC's 1988 Proposal Call. This document sets out a number of abnormal specifications for the construction of a road tunnel under the Strait. The width of a tunnel, it specified, was to be 14m (45 feet), a dimension that exceeds the capacity of current tunnel-boring technology. To build a tunnel of such a width would require blasting, a much more costly construction method. Tom Kierans argues that these specifications made it inevitable that a road tunnel could neither be built nor, due to the necessity of costly ventilation, operated economically.

Not surprisingly, PWC received no response to their single-tunnel proposal call. But it did receive a proposal for a three-bore road tunnel. The reluctance of developers to make a proposal on a construction project that is unnecessarily expensive to build and extraordinarily expensive to operate is understandable. Not understandable, however, is why PWC would set out such specifications, and why the only tunnel proposal received was then rejected. PWC's explanation? "A tunnel is not a viable response to the requirements of the Proposal Call."

Under the heading "Acts of Omission" is the fact that the technical data which PWC provided to potential developers lacked the information necessary for putting together a viable tunnel proposal. Geoconsult, PWC's tunneling advisors, warned it in 1987 that its data was inadequate, saying that it was "necessary to make some additional investigations to confirm or verify the rock-physical and hydrographical data which are available so far." It added that "we strongly recommend an excavation for a smaller tunnel in advance" in order to acquire the technical information that would be necessary for any developers to put together a reasonable tunnel proposal. PWC ignored this advice, reasoning, no doubt, even then, that "a tunnel is not a viable response to the requirements of the Proposal Call."

Tom Kierans is also troubled by PWC's refusal to compose, in response to the EAP's request, "an easily comprehensible diagram" that would illustrate the comparative risks and benefits of the bridge and tunnel options. PWC argues that this was an impossible task, but Kierans claims that "such diagrams are normal for all major projects." In consultation with other engineers, he drew up just such a diagram which showed that a bridge had a significantly poorer rating than did either a road or rail tunnel. Of PWC's obstinacy, Kierans says, "I personally do not think that you should enter into any consideration on any professional basis of any project unless you can make some kind of a comparison as to what the benefits and the risks are."

The three-bore road tunnel was submitted by a group that included the Quebec-based Lavalin Corporation, which was also a partner in the 1985 Omni System group that proposed a rail tunnel. Lavalin Vice-President Armand Couture addressed the EAP in March 1990. He, like Tom Kierans, criticized PWC's refusal to make cost-benefit comparisons among the various options, and argued that the technical and environmental data and the costs of other fixed link possibilities should have been given to developers at the same time such information relating to a bridge was made available. Comparisons, he said, should have been made of all the possible options before a single one was selected, especially for a project of this magnitude. He added that, in terms of cost, a bridge and the three-bore tunnel proposal were "of the same order of magnitude." However, he continued, the price tag for the Omni System rail proposal, already rejected by PWC, would have been "slightly above 50 percent" of the other solutions.

Attending that same 1990 EAP meeting was R. W. Harmer, a Director of Bechtel Canada, subsidiary of the American-owned Bechtel Group, one of the world's largest engineering firms. Although his company had submitted a bridge proposal, his address to the Panel was on the subject of tunnels. He also complained of the pro-bridge bias of the geological data PWC made available to developers, and added that, given the knowledge and experience gained by the construction of a tunnel under the English Channel — the "Eurotunnel" or "Chunnel" — that project "provides really a blueprint of proven ideas for incorporation into a crossing for the Northumberland Strait." Harmer told the EAP that, because the available geological information was suitable only for bridge construction, PWC should contract for a geo-technical study for a tunnel, and that to build a tunnel to PWC's unrealistic specifications would cost one-and-a-half times as much as a bridge. But, he added, the cost of a single-bore rail tunnel "would be quite a bit less than what our bridge estimate was," and he felt that it would "be prudent at this time to consider a single-rail tunnel using an electrically powered train."

Two years later, in an article in the *Eastern Graphic*, E. van Walsum made much the same point when he said, "... the cost of building, maintaining and operating a single track railway tunnel to achieve a design life of 100 years can be expected to be many millions of dollars less than the similar cost for a two-lane highway bridge with the same design life of 100 years. A bridge would be exposed to the highly corrosive and physically aggressive ocean environment. The costs necessary to maintain the bridge would, in the end, be paid by Canada and the travellers to and from the Island."

Walsum's comments are especially interesting because of his experience with the PEI fixed link. A civil engineer from Pointe Claire, Quebec, Walsum served as a bridge design engineer with the government of Alberta, and in the late 1950s, as project engineer with the consulting engineering firm FENCO, responsible for the design and construction supervision of the Princess Margaret bridge in Fredericton and the Hugh John Fleming bridge in Hartland, both over the Saint John River in New Brunswick.

In 1963, as a partner in a consulting firm, he reported on the basic layout and cost of four tunnel schemes which, together with a bridge and a causeway, were to constitute the fixed link between New Brunswick and PEI. The tunnel alternatives he was asked to study were for two-lane highway traffic and for two-lane highway traffic plus a single track railway.

The scheme chosen by PWC, upon which construction was started and subsequently discontinued, would have provided for both highway and railway traffic. The question "Why not just a railway tunnel?" was put to PWC at that time, Walsum said, but was never satisfactorily answered.

In 1989, the EAP again asked that question of PWC, and again a satisfactory response was never provided. From the beginning, PWC excluded a rail tunnel from consideration, narrowing the choice to the bridge or highway tunnel options. But the EAP struggled to understand why even a highway tunnel was ruled out and, in 1989, demanded that PWC explain why it was so adamantly opposed to a highway tunnel. The response came in a seven-page letter, written in bureaucratic idiom and signed by Project Manager Jim Feltham, in which he offered a total of two reasons for the rejection of a road tunnel. One of these was a direct consequence of failing to drill a test tunnel, as was recommended to PWC by its own Austrian consulting firm in 1987. The other reason Feltham gave was "on the grounds of regional benefits."

Feltham's mention of "regional benefits" may seem bland, but behind it lies one of the most glaring flaws in PWC's handling of the entire project. One condition of PWC's 1988 call for proposals for construction of a fixed link — entitled, innocuously enough, "Addendum No. 1" — concerns "Regional Benefits Requirements." Here, the employment and purchasing that the various proposals would engender are evaluated. If something is purchased in one of the "Special Designated" areas of the region, it is given a "weight" of 4: if it is purchased elsewhere, a descending scale assigns it a smaller value. (For example, if a developer planned to purchase structural steel which amounted to 3% of the total project cost in a "Special Designated" area, PWC would assign it an "Index Value" of 12, by multiplying 3% by 4.) Similar regionally based calculations were made for labour.

By using a percentage of a proposal's costs rather than a percentage of its *needs*, PWC strongly biased its decision in favour of proposals that require the most labour and materials. The Report of the EAP says that a bridge would require 5,000-7,000 person-years of employment during construction, a tunnel only 1,000. A tunnel, requiring fewer workers would receive a lower evaluation in the area of "Regional Benefits Requirements."

This bizarre, more-is-better thinking exposes the logic of the tunnel specifications in PWC's Proposal Call. Those specifications ensured that a tunnel would never be built, simply because tunnel construction requires only a fraction of the number of jobs of bridge construction. PWC's assertion that "a tunnel is not a viable response to the requirements of the Proposal Call" now makes sense: It is not a "viable response" because PWC never intended it to be. (For PEI, however, tunnel construction would mean that Islanders would receive a much larger slice of the "regional benefits" pie — tunnel construction could proceed from both sides at once.)

PWC and local boosters have sought public support for their bridge project by hailing the "regional benefits" it would bring. In doing so, they have turned the whole question of a fixed link from one that asks, "What's best for Prince Edward Islanders?" into one that asks, "What's in it for me?" Brazen individual self-interest has replaced consideration of the future of Island society.

In a speech to the Progressive Conservative Party's national convention, then-Prime Minister Brian Mulroney bragged about the decisiveness of his government. It was decisive, he said, even in the face of opposition from the Canadian people, and he went on to give a few examples. Free trade and the GST were atop his list. He didn't say whether the policy decisions were correct, only that they were decisively made.

This decisiveness-as-divorced-from-popular-will has permeated all levels of the Tory government. In relation to PEI and the fixed link, we see it in the form of denial — denial that the method selected to span the Strait, a bridge, might be faulty, and denial that there can be another way, a better way, an alternative.

The fixed link project has been reduced to a make work project and is being sold to Islanders on the basis of the number of jobs it will create. Only a fraction of this work will go to Islanders. These temporary jobs, along with the 60 to 80 permanent ones that will be created by the operation of a bridge and an unknown number of low-paying, seasonal tourism-related jobs must be considered against the 651 well-paid and permanent jobs currently provided by Marine Atlantic, the jobs provided by the Wood Islands ferry service and the spin-off employment that comes from the almost $25 million from wages, benefits and purchases that Marine Atlantic pumps into the Island economy every year. As well, the potential damage that a bridge might bring to the $75 million annual Northumberland Strait fishery has to be weighed in the balance.

And the scales that Islanders use to do that weighing should not be borrowed from local Tory politicians or PWC. That scale should be Islanders' own.

The reasons for PWC's bias, tunnelling expert Tom Kierans said in a letter to the Globe and Mail, "have never been satisfactorily explained." He sums up PWC's preference for a bridge this way: "The big problem was the single-minded bias against the rail tunnel and, almost the other way, the single-minded determination to build a bridge, whether you liked it or whether you didn't ... They are determined to build a bridge, no question whatever ... I have never seen anything as single-minded or as biased for a bridge as I saw in that bidding procedure."

Kierans' assertion that PWC's bias has "never been satisfactorily explained" is correct. But the reasons for it are, nonetheless, explicable.

When a rail tunnel was ruled out by the Ottawa Tories back in 1986-87, the most sound option for a fixed link was buried. This decision was not made on the basis of what was best for Islanders. It was not made on the basis of what was best for taxpayers. It was not made on the basis of what was best for the environment. It was made, quite simply, because such a decision best fit the federal Tories' corporate agenda of privatization and deregulation of transportation. The remains of the fixed link idea were then given to local elite to scavenge on — and scavenge they did.

Decisions born of such a process are devoid of reason, responsibility and any desire to ensure the best for Prince Edward Island. As of this writing, PWC is on the verge of signing a contract for the building of a bridge across the Strait. Islanders owe it to themselves, to endangered ferry workers and to those involved in the Northumberland Strait fishery, to call a halt to this most ill-conceived and devious of projects. Islanders, although possibly too exhausted with fixed link discussions to immediately force a reopening of investigations into the *best* crossing option, will then, at the very least, have defeated this current fiasco-in-progress.

Chronology of the Fixed Link

(with Editorial Comment from the *Eastern Graphic*)

1832 Steamer service begins between Pictou and Charlottetown.

1877 "Northern Light" steamer establishes a regular winter connection to the mainland except in heavy ice and storms.

1885 Senator George Howlan proposes constructing a tunnel under the Strait.

1956 PEI government approaches the federal government with proposal to investigate feasibility of a permanent crossing.

1958 Consulting engineers and government agencies determine that a rock-filled causeway is feasible but that the effects of ice and tides would require attention.

1965 Federal government decides to proceed with the design and construction of a causeway, bridge and tunnel crossing for road and rail.

1969 Plans for proposed crossing are abandoned when the province opts instead for an economic development agreement and improved ferry service.

1985 Three unsolicited proposals to design, finance, construct and operate a fixed crossing are received by the federal government.

Three proposals have now been made for a permanent crossing of the Northumberland Strait. Environment Minister Tom McMillan has been pushing the idea of a permanent crossing in the past year. He says that $200,000 will be spent on a feasibility study of the three proposals in the next few months.

The cost figure for the crossing could be over $500 million and it is being claimed that this will come from private sources rather than from public funds. (*MacN*[1] 20 Aug 86)

Is PEI going to have a fixed crossing to the Mainland?

If one were to listen to Environment Minister Tom McMillan we'll have one in the next few years. In fact, he's even telling us now that the crossing has been "fast tracked" by Ottawa.

"Fast tracking" is the term used by politicians when they want things done in a hurry. This usually means that people have little say in the decision. (*MacN* 29 Oct 86)

June 1987 Twelve consortia respond to expression of interest request.

16 Nov 1987 Stewart McInnes says the federal government will be calling for tender proposals.

20 Nov-1 Dec 1987 Public Works Canada (PWC) sponsors open houses on PEI.

Prince Edward Islanders are getting a fixed strait crossing whether they want one or not. The proposal is being hustled by the federal government and the province is simply standing by doing nothing ... Islanders are only being given two weeks to make up their minds about the crossing. They haven't even been given the full details of all the studies already carried out by federally financed consultants. The eleven days of information are simply a series of Open Houses at which the federal view will be presented.

There is no real opportunity for Islanders to question what is being proposed. There is no real opportunity for those who are opposed to mount any substantial opposition because of the time frame ... The federal government has had months and months to promote its position. Why must Prince Edward Islanders be given so little time to respond? ... The point is that the federal government wants to give the impression that it is consulting with the public but the reality is that it isn't. People need time to study something as important as this. (*MacN* 18 Nov 87)

22 November 1987 Friends of the Island hold initial meeting.

7 December 1987 Writ of Plebiscite signed.

How is the ordinary Islander going to get through all of this information that technical experts have taken ten volumes to compile? And those are just the assessments favouring the fixed crossing. What about the arguments against?

This is precisely where the federal government's own Environmental Assessment and Review Process (EARP) is needed. It gives time for people to decide. It gives time for the creation of an independent panel to investigate all the issues. It gives time for information to be given to the people who will be affected. (*MacN* 9 Dec 87)

9 December 1987 Friends of the Island rally at Charlottetown Hotel.

17 December 1987 Islanders for a Better Tomorrow Inc. is incorporated.

18 December 1987 CBC television debate on the fixed link.

One of the panelists on last week's TV debate on the Northumberland Strait fixed crossing said that it was time that people put their trust in the government, in the bureaucrats and in those who have studied the need for the fixed link.

He was speaking for the YES side.

He's right. It is a question of trust but that is precisely why Islanders should vote against the fixed crossing.

How can we trust Public Works Canada when it attempted to hide two important studies on the economic feasibility and financing of the fixed link? ...

Are we to trust Environment Minister Tom McMillan when he's the major promoter of the crossing? ...

Are we to trust a federal government that suddenly tried to buy off Islanders who fear what effect a fixed crossing will have on the Wood Island's ferry? ...

Are we to trust consultant's reports that constantly admit that there are gaps in the information available? ...

Are we to trust the future of the fishing industry simply to an assurance that compensation will be paid? ...

Are we to trust politicians who glibly tell us that the loss of 651 ferry jobs will be softened through early retirements and with re-training schemes? ...

Are we to believe businessmen who unhesitatingly promise lower transportation costs but who are hard pressed to prove the savings when other factors like bridge tolls that are higher than ferry rates come into effect? ...

This is far too important to be simply an election issue. There are too many unanswered questions; too little information; too many effects that have important repercussions ... (*MacN* 23 Dec 87)

14 Dec 87 - 16 Jan 88 The Institute of Island Studies holds Public Forums.

The present meetings are being held only because of the provincial government's instructions. Public Works Canada would have been quite happy to have finalized it all with the last of its open houses on December 1, 1987.

Legitimate inquiries for copies of the consultant's reports were practically ignored by the project office. This newspaper managed to get reports earlier than others but that was almost despite the efforts of Public Works Canada. Others weren't so fortunate ...

None of the news media did a completely effective job in informing readers, listeners or viewers. Studying those reports took a lot of time and effort. Even further effort was needed to put the information into an understandable form. Yet it should have been done. Polls and phone-ins aren't a substitute for plain simple facts by reporters.

The news media or at least the press should also have been pointing out the inadequacies of the consultant's reports. There are conclusions that can't be substantiated. There are areas of concern that haven't even been looked at. There's inadequate data. Some of the reports were completed in too short a time frame ...

If people have questions now, what will they have when a government is given a Yes vote? (*MacN* 6 Jan 88)

18 January 1988 Plebiscite held: 59.46% vote Yes, 40.21% vote No.

I never really thought of myself as a potential subversive when I voted No in the plebiscite. The reason I voted No was because Minister Tom and the consultants didn't convince me that they had looked very thoroughly at the environmental effects of a crossing, especially if it turned out to be a bridge.

I liked what Premier Ghiz said that night on the television ... that he might call for an [Environmental Assessment Panel] EAP. But then I read the next day where Minister Tom said people who wanted an EAP to make sure a bridge wouldn't muck up the strait fishery were "subversive."

It's funny you know, you can't really tell who is a subversive just by looking at them. Some of them wear three-piece suits as a disguise and seem to mix in with the population without any trouble at all. The only thing they have in common is this peculiar belief that there really isn't a big rush to build the crossing if it takes a few months to really examine all the effects it might have on the Island.

But that's the deceptive way we subversives think. It takes the trained mind of Minister Tom to see through us in a minute and identify us for what we really are. We are people who don't have our priorities strait. (*McA*^{2} 27 Jan 88)

Islanders aren't going to get a real environmental assessment of the Strait fixed link. It's obvious the federal government doesn't want one and won't approve of one easily ...

Many people feel strongly that the only way that the whole issue can be studied fairly and impartially is to set up an EAP. McMillan doesn't want one. McInnes refuses to ask for one. Ghiz waffles and says a committee might ask for one.

And Islander's concerns are going to be passed over lightly as the federal government moves to get the project going as quickly as possible ... (*MacN* 27 Jan 88)

Jan-Dec 1988 Stewart McInnes refuses to call for EAP.

Public Works Canada has been reluctant to move in any direction except its own fast track with regard to the Strait project ... Stewart McInnes has stated that "I will give Premier Ghiz the ammunition he needs to say to the Island people that a fixed link is environmentally safe."

It isn't ammunition that's needed. It is facts. (*MacN* 3 Feb 88)

March 1988 Generic Initial Environmental Evaluation (GIEE) is published. Call for proposals is issued.

Last December and January when Islanders were discussing the whole question of a fixed link as they headed for the plebiscite they were never told that the initial environmental evaluation was "produced in considerable haste last fall (1987) and there were a number of errors and omissions we wish we had found before it was distributed." The newer version of this evaluation was published in March, 1988, and it was only then that Islanders could find out that one of the main documents for the whole plebiscite debate contained many errors and omissions.

As far back as lst March officials with Public Works Canada knew that the tunnel proposal wasn't even being given any serious consideration but this was never published. Yet they knew that Island fishermen and others had voted Yes in the plebiscite because they felt that a tunnel was the best alternative. (*MacN* 19 Oct 1988)

14 June 1988 Eight fixed link proposals are unveiled in Charlottetown.

Public Works Canada is hell bent on pushing through with the evaluation of the fixed link to the mainland in as short a period as possible. It is determined to take the same course it did last December when it allowed a few short weeks for the public information process ...

It is using precisely the same tactics with the evaluation process. Public Works Minister Stewart McInnes has refused to release the particulars of the eight different proposals, seven for bridges and only one for a tunnel. The general public is given only the most rudimentary details of the eight projects and no opportunity to have any input into the evaluation ...

One of the most disturbing aspects of the evaluation is that it will be based solely on the lowest price. Where does that leave the environmental aspects? Or what about the impact of a fixed link on the fisheries. (*MacN* 22 June 88)

30 September 1988 Three finalists are announced in a press conference by Stewart McInnes, and all are bridge designs. Federal Environment Minister Tom McMillan and Premier Joe Ghiz do not attend the press conference.

1 October 1988 Federal election is called for November 21, 1988.

It isn't just environmentalists that have been seeking a panel review process for the bridge crossing. It also includes fishermen, truckers, farmers, some businessmen and many other ordinary Islanders. Several national groups have also sought the same thing including advisors to Environment Minister Tom McMillan ...

Up to now Stewart McInnes hasn't gotten the message. Neither has Tom McMillan. But with a federal election now underway, they may suddenly see that this is really their chance to give Islanders what they want — a chance to examine the whole project openly and without pressure. (*MacN* 5 Oct 88)

McMillan's response to his critics in this election has been one of the most astounding factors at the local level. He's accused Dr. David Suzuki, a nationally recognized broadcaster and scientist, of distortion in his statements and Elizabeth May, an environmentalist who quit as his policy adviser, of blackmail. Those are serious accusations ...

McMillan's outbursts against May and Suzuki in the campaign shouldn't really come as a surprise to Islanders who have been subjected to McMillan's outlandish claims themselves. Before last January's plebiscite he suggested that Islanders opposed to the fixed link were "ignorant." Later he said that those wanting further environmental studies were acting as "subversives" to the fixed link ...

Elections aren't usually won. They're lost. Tom McMillan may well lose his seat in Hillsborough but if he does, it's his own fault. (*MacN* 16 Nov 88)

21 November 1988 Federal election. PCs Tom McMillan of Charlottetown (Minister of the Environment) and Stewart McInnes of Halifax (Minister of Public Works) lose their seats. Nationwide, Conservatives score a landslide victory.

12 January 1989 In a surprise move, the federal government announces its decision to refer fixed link project to an EAP.

It passed almost totally unnoticed in all the press coverage of the delay in the fixed link decision while a full EAP is held ... It was the quote from Glenn Duncan to the effect that the Public Works Canada would appear before the EAP as a proponent of the project. A proponent my dictionary tells me, is a fancy word for ... "advocate, backer, enthusiast, patron, promoter, endorser, propagandist ..."

Well, ain't that just dandy ... good old Glenn and Public Works Canada have finally come out of the closet ... They have blown their cover and we can now evaluate the impartial quality of their information in a new and very different light ...

The fact that Glenn and his department have been playing this game with us for two years is bad enough. The fact that they have been commissioning, interpreting, laundering and releasing all the available information on the project amounts to a totally cynical view of the concerns Islanders might have about the project and its effect on their homeland. We never have been given any kind of explanation as to why the tunnel was a non-starter, except that it didn't meet departmental criteria. No wonder, if the department and the federal government rigged the criteria in the first place ... (*McA* 25 Jan 89)

28 April 1989 Environmental Assessment Panel is appointed.

May-June 1989, March 1990 EAP hearings.

The Environmental Assessment Panel reviewing the Northumberland Strait fixed crossing says that it requires additional information on the project from Public Works Canada ...

The Panel's decision obviously confirms what many people have been saying about the Public Works Canada studies. They are not adequate. A lot more information is needed. The tunnel option is still open for discussion as far as the Panel is concerned. It didn't accept Public Works Canada's position on the tunnel alternative ...

One of the most astounding things with regard to the fixed link is the insistence that the submissions by the developers are private and that the public has no right to them ...

What is even worse is that they aren't available to the Panel review — the very agency set up to ensure that adequate protection and safe-guards will be in place if the project is built ...

It isn't the bureaucrats of Public Works Canada that will be affected by a fixed link. It will be Islanders. (*MacN* 30 Aug 89)

The EAP decided last week that it would go into the final round of public hearings this March, even though Public Works Canada still hasn't satisfied the members on several key points ... Consultants were retained by the Panel to take a look at the adequacy of the PWC response to the questions asked. They do not give PWC good grades, especially on absolutely vital points. Indeed, they raise enough questions of their own to cast doubt on the integrity of just about everything PWC has done ... (*McA* 28 Feb 90)

The only moment of any real drama at the fixed link hearings in Charlottetown last week comes during the session reserved for discussion of the tunnel option. The Panel's tunnel expert, Thomas Kierans from St. John's, asks several embarrassing questions of James Feltham of Public Works Canada, to which he gets somewhat misleading replies.

The audience has their chance for questions, and then Carol Livingstone, from up west, leans forward, hits her microphone button and says ... "I just want to get this straight, Mr. Feltham. Am I to understand that with all of these potential problems with a bridge, you people dismissed a rail tunnel out of hand, just because people might have to wait a few minutes to get across the Strait?"

James Feltham, the Public Works Canada point man on the fixed link project, punches up his own microphone and gives a short and definitive answer. He says ... "yes" ... and that one word answer speaks volumes in its own simple eloquence. It tells us we were hoodwinked by Public Works Canada when we voted two years ago ...

A tunnel under the Strait never was a contender as far as Public Works Canada was concerned. A rail tunnel (that's one where you drive on a flat car to be transported through the tunnel) was tossed out before the environmental process started. A drive through tunnel was dumped by a more intriguing gambit ... (*McA* 4 April 90)

15 August 1990 EAP Report is released. It recommends that the bridge project not proceed.

Ultimately, the Panel looking into the matter of a fixed link to the Mainland agree[s]. The bridge concept, as proposed, was unacceptable ... a bad idea. It was also a bad idea badly put by the federal Public Works Canada. Whether through incompetence or over confidence, or perhaps simply because the case couldn't be proven in the way it was attempted, PWC must carry a large measure of responsibility for the Panel's decision. Indeed people like Jim Larkin, of "Islanders for a Better Tomorrow" and Gary MacLeod of the Chamber of Commerce, could well redirect their resentment from ferry workers and fishermen and others who disagreed with them, to PWC where the blame rightly belongs ...

The pro-linkers have also criticized the Panel for considering the plight of ferry workers and fishermen in their deliberations, and factoring in the economic losses the fishery and the ferry workers' payroll would pose. It is as though people did not count in their lexicon, or at least only the people the pro-link groups represent, chiefly those who think they would benefit in some economic way. The pro-linkers seem quite willing still, to throw their fellow Islanders to chance and fate. They forget to factor those losses into their "better tomorrow" ...

Rather than doggedly cling to the dream of a bridge across the Strait, the pro-linkers might do themselves and their province more good by expending their energies in a joint effort to get that better ferry service. It is the one alternative that no one would disagree with. It will unify Islanders instead of dividing them. It will benefit all of us without penalizing anyone. (*McA* 22 Aug 90)

It's funny how viewpoints change. Take the case of the need for an Environmental Assessment Panel review of the fixed link to the mainland ... The Panel review was held. Nobody objected to the six members on it. Nobody objected to their line of questioning. Nobody objected when they asked for more information from Public Works Canada ...

Now a review Panel has rejected the Northumberland bridge proposal on environmental and socio-economic grounds. Nobody objected to the Panel or how it conducted its work before the release of its report. Now the Panel is being questioned by some. Its findings are questioned by others, while others are calling for a new Panel review of a bridge proposal Islanders couldn't even find out about. It seems to be a case of trying to change the rules until a suitable finding results. (*MacN* 29 Aug 1990)

24 September 1990 Premier Joseph Ghiz writes to PWC with suggestions on how the bridge project could go ahead despite EAP recommendation that it not proceed.

November 1990 Federal government reacts to the EAP Report. An ice panel is to be appointed.

The [Environmental Assessment] Panel made its decision on the fixed link quite clear. It was against the construction of a bridge because of environmental aspects. The main one was that a bridge would delay the ice movement in the Strait by one or two extra weeks ... Now according to Elmer MacKay, there's to be a new panel of ice experts to say whether a bridge is environmentally safe or not. And do it by next March too! There's going to be a new ice model apparently looked at by new ice experts. But does that change a thing? There aren't going to be any more studies of real ice conditions in the Strait. Just computer models. What does that tell you? ...

It's the same old figures being re-done. Islanders have been told they're environmentally unacceptable now. Why should they accept a new version of them after next March? (*MacN* 28 Nov 1990)

26 June 1991 Proposal call to developers.

The whole issue of the fixed link is getting even more curious by the week. When the Environmental Assessment Panel rejected the idea of a bridge across the Northumberland Strait on a number of environmental issues, some people thought that was the end of it. Far from it! Public Works Canada has continued full speed ahead as if the EAP rejection never existed.

PWC chose to ignore all the issues raised by the EAP except for the ice-out one ... PWC's plan to get around this was simple. Appoint a new "ice committee" and get a new assessment ... Surprise! Surprise! The new ice committee was able to give the go ahead. (*MacN* 14 Aug 91)

31 January 1992 On to the pricing stage.

27 February 1992 Islanders for a Free Ferry Service announce the formation of their new group at news conference.

Make no mistake about it, a fixed link to the mainland in the form of a bridge, is on a fast track to completion ... [Last week], Federal Finance Minister Don Mazankowski, speaking in Ottawa ... used the bridge as the kind of private initiative mega-project the feds want to encourage, he signalled that the bridge had passed muster at cabinet level, and even more importantly, at the deliberations of the Tory election committee. That's the real reason for the fast track. It has little to do with the lives of those of us who live here. It has everything to do with a political party desperately looking for a massive job creation project to lift the region out of the current depression in time for the election soon to come. (*McA* 5 Feb 92)

27 May 1992 Subsidy bids of the 3 developers are opened. No submission fully complies with the proposal call.

17 July 1992 PWC announces initiation of discussions with lowest bidder to submit plan acceptable to federal government.

Memo to Elmer MacKay: Points to stress when you present the fixed link proposal to cabinet shortly. Public Works Canada employees and its hired consultants have downplayed safety warnings no matter who issued them. A number of the warnings are well documented but still ignored. You will recall that similar warnings were issued prior to the opening of the Westray Mine. You, along with other politicians and government employees, chose to ignore those safety warnings. You and your colleagues won't want to have the same kind of regrets, I'm sure ... Mr. MacKay, I know you're a busy cabinet minister but I know that you'll want to consider this project very carefully. After all, neither you nor I nor anyone else wants another Westray disaster. (*MacN* 30 Sept 92)

7 November 1992 Coalition Against the Bridge protest/rally held in Borden.

16 Dec 1992 Joe Ghiz, Elmer MacKay and Frank McKenna sign a political agreement to proceed with the bridge project.

Well, well, well! Paul Giannelia of Strait Crossing Inc. [SCI] also found it difficult to get information about the fixed link ... What about Islanders? We're the ones that haven't been getting the information. Not Mr. Giannelia. He's the one that beat out seven other proposals. We don't even know what was in those proposals. We don't know why they were rejected. We don't know anything about them. For that matter, there is very little known even about SCI's proposal. (MacN 21 Dec 92)

... the bridge believers are the true romantics of this island. Their unshakeable conviction is unhampered by any specific evidence ... The reality is, it's a risk and a gamble from an engineering, environmental, social and economic perspective. Never mind all the millions spent on consultants, the truth is we just don't know, and we never will know until it happens, what the effects will be for good or ill. (*McA* 30 Dec 92)

January 1993 SCI holds 9 public meetings to explain its Environmental Management Plan.

Paul Giannelia has been a smooth spokesman for the Northumberland Strait Bridge project. Low key, knowledgeable, gentlemanly and accommodating when addressing the public meetings held by his company, Strait Crossing Inc. ... Mr. Giannelia has made a number of promises during the course of the public meetings. Here they are ... :

He promised to "Get an icebreaker and run it up and down the Strait." ...

He promised SCI would pay more than the $10 million committed for compensation to fishermen if necessary ...

He promised ferry service if the bridge is closed ...

He promised that not a day, not an hour of time would be lost by motorists using the bridge. He obviously hasn't experienced just an ordinary winter snowstorm in the Strait ... (*MacN* 20 January 93)

30 January 1993 The *Guardian* announces opening of SCI office at 10 Pownal Street in Charlottetown.

... New material had to be filed by Public Works as part of the government defence against the court action launched against the feds by the Friends of the Island and their friends. So the new study of the other studies became part of the public record ... [It concluded] that the critics were absolutely correct — the economic benefits of a bridge rather than a ferry service remains an unproven proposition ... They agree ... there is no proven long term economic benefit to a bridge over the existing ferry service ...

If reason was guiding this mega-project, it never would have gotten this far. But since reason has nothing to do with it, do not expect that this devastating assessment will slow the process in the slightest ...

The reason it doesn't matter, is that this project has never really had very much to do with the people of Prince Edward Island anyway. So whether it's three strikes or twenty against the idea that the bridge will be of economic benefit, that doesn't count a spoonful of spit. All that matters is that fools fly in, when reason takes flight. (*McA* 24 Feb 93)

1 March 1993 Friends of the Island court challenge to the fixed link begins in Toronto.

19 March 1993 Justice Barbara Reed of the Federal Court of Appeal hands down a 56-page judgement. The decision states that further environmental assessment of the specific bridge plan is needed before the project can proceed: "It is particularly disturbing, in this case, to find that a generic design was referred to a Panel when the government had access to more detailed information, respecting the three concept proposals being considered, which was not referred ...

It seems downright silly and an incredible waste of public funds and people's time to find, at the end of the day, that one of the reasons for which the government rejects the Panel's recommendations is that the Panel did not have before it the detailed information to which the government was privy and which the government had refused to provide to the Panel."

Public Works Canada is still at it. Trying to have us believe it has lived up to its responsibility with regard to the Northumberland Strait bridge.

The department spokesman last weekend claimed there wouldn't be much delay because of Federal Justice Barbara Reed's decision ...

Wrong. (*MacN* 7 April 93)

Notes

1. Jim MacNeill, editor of the *Eastern Graphic*.
2. Jack McAndrew, editorialist with the *Eastern Graphic*.

Economics

"After the thrill of expansion is gone and after the rumble of the big trucks are silent; after the newly built roads have grown up in alders again; after the line-ups at the take-outs are gone; after the fat paychecks are gone and when the bars are no longer crowded and you know almost everyone in them; when the cars going by contain men wearing baseball caps instead of hard hats and suits; when the work camps are quiet and decrepit and run down; when the entrepreneurs are back to square one; when Cape Tormentine is a ghost town and Cape Jourimain is a string of last chance gas stations; what then?"

Steve Jones (fisher),
Environmental Assessment Panel (EAP) hearings,
15 March 1990.

From the beginning of the fixed-link-fandango, Canadians have been assured that any project to which Public Works Canada (PWC) would put its name had to jibe with five evaluation criteria. Two of those criteria were that the project had to be financially viable and economically beneficial. Evaluation working groups, PWC assured us, "were comprised of various technical consultants and representatives from provincial departments as well as participants from all relevant federal agencies." Thus assured, we were not to worry our heads — we were in good hands.

But some people refused to be cosseted and did worry about the financial viability and the economic benefits of which PWC spoke so grandly. And there developed quite a literature of arguments refuting the cock-eyed economics of a project that would put an end to 651 permanent unionized jobs in a province where, in February 1993, 20.9% of its workforce was unemployed. Even studies done for bridge promotors had an uphill battle fingering any economic benefits which would outlive the construction phase of the project.

Looking towards the other end of the project, a former Islander witnessing the pilons of a former Hillsborough Bridge decaying into the Hillsborough River at Charlottetown asked the costly question: "If we can't affort to clear up these few piles of rubble, who's going to pay to take down the Northumberland Strait bridge?"

PWC's economic feasibility assessment — the study on which it based its decision that it was okay to go ahead with the bridge — has been thoroughly discredited. Initially, this was done in a piece-meal fashion, but now a thorough re-assessment from a number of sources confirms that the study "does not establish the economic viability of the fixed crossing." Particularly worrisome to those who had faith in the integrity of PWC is the department's attempt to bury the study that came to this conclusion. The study only saw the light of day because legal proceedings forced it to be made public.

Financially, too, the project is worrisome: When is a subsidy not a subsidy and when is it a *carte blanche* dive into taxpayers shrinking pockets? The loading up of one Strait crossing option with benefits and inflation-indexed tolls and transfers, while the current crossing option — that one which is economically viable — is starved of equipment and any kind of inflation index except one associated with costs, is yet another not-too-subtle manipulation of the dance.

The Economic Feasibility Assessment of the Northumberland Strait Crossing: A Critique

Peter G.C. Townley

The controversy surrounding the proposed fixed link project between Prince Edward Island and New Brunswick has centred largely on questions concerning a bridge's potential environmental impacts and how any type of fixed link would alter the "way of life" of Islanders. These contentious issues have received much media attention, while the basic matter of whether or not a fixed link is economically viable has been overshadowed.

Economic viability would require that, during its economic life, the social benefits of the project exceed its social costs. In 1987, Public Works Canada (PWC) commissioned Fiander-Good Associates Ltd. (FGA) of Fredericton to perform a cost-benefit analysis of this project. FGA conclude:

> In summary, the economic results indicate that a fixed crossing, more specifically a fixed crossing involving a bridge, will be similar to the existing ferry services with respect to the net present value of society costs. A lower discount rate [than 10 per cent] would favour the fixed crossing alternatives. (FGA, 1987:2.3).

Although we have more to say on the matter of discount rates later, one notes that the Treasury Board's 1976 "Benefit-Cost Analysis Guide" recommends a discount rate of 10%. Based on this, FGA's conclusion means that replacing the ferry service with a bridge would neither enhance nor diminish society's well-being, although it would seem that the above was interpreted by PWC to mean that a fixed link in the form of a bridge is economically viable.

The FGA study, however, is seriously flawed. The methodology used is biased significantly in favour of bridge construction and neither FGA's conclusion nor PWC's interpretation of it is warranted. In this essay, an updated and expanded version of Townley (1992), we outline the FGA study and explain why the methodology used is incorrect. We also note errors and omissions which lend to this bias. As the FGA study does not provide enough information to allow us to redo the analysis completely and correctly, our modest conclusion is that, at the very least, it would be inappropriate to begin construction of a bridge until it has been subjected to a more meticulous economic analysis.

The Fiander-Good Study

At the outset of their report, FGA make clear what PWC required of them, by reproducing the terms of reference for their analysis:

> The economic feasibility study is to be of the benefit-cost type wherein all capital costs, public facility maintenance and operating costs and economic development costs are to be expressed in present value terms for each alternative, including the sensitivity to changes in assumptions. The Stanford and the DRIE-TC-PWC analyses are of the type expected. (FGA, 1987:1.2).

Unfortunately, FGA do not provide a bibliographical entry for either a "Stanford" or a "Department of Regional Industrial Expansion-Transport Canada-Public Works Canada" study, although we know that Transport Canada did commission the Stanford Research Institute to conduct a cost-benefit analysis of a fixed crossing in 1968. Indeed, by then construction of a combined causeway-bridge-tunnel had already begun, but was terminated in 1969 before much was accomplished (Perchanok, 1988:17-18). As we have been unable to obtain a copy of this 1968 report, we cannot determine if the errors documented below are the result of FGA misapplying the methodology of the 1968 analysis or if it is the methodology of that earlier report which is incorrect. Given the questions FGA themselves raise regarding the methodology they eventually used, the latter appears to be the most likely circumstance.

Because provision of year-round communication between PEI and the mainland was a condition of the Island joining Canada in 1873, FGA's task was to examine what kind of transportation service should be provided, not whether any should be provided at all. Their study is thus, quite properly, of the cost-effectiveness type of cost-benefit analysis, and the five alternatives they examine are:

1. The status quo (the existing ferry services);

2. A bridge replacing both the Borden-Tormentine and Caribou-Wood Islands ferry services;

3. A bridge replacing the Borden-Tormentine ferry service, but retaining the Caribou-Wood Islands ferry service;

4. A rail tunnel replacing both ferry services; and,

5. A rail tunnel replacing the Borden-Tormentine ferry service, but retaining the Caribou-Wood Islands ferry service.

Although bridge and rail tunnel options were assumed to have an economic life of 100 years, the study period was limited to 40 years — 1987-2027 — and the "... continuation of the existing ferry services and the associated costs were considered the base case from which the four remaining options were compared with respect to their present worth costs and incremental benefits." (FGA, 1987:1.2). Their objective, then, was to identify the transportation alternative which would minimize net social costs.

The categories of benefits attributable to a fixed link examined by FGA were travel time savings, corporate income taxes raised, reduced vehicle operating costs, salvage value of the infrastructure and economic development.

Concerning this last item, FGA attributed zero developmental benefits to a fixed link because the main impact would be on tourism and "... a tourism gain by one province generally results in a loss by another province, thus resulting in benefits being netted out from a global perspective." (FGA, 1987:5.17). This view is echoed in a later report prepared for PWC. In their investigation into the impact of a fixed link on tourism, Smith Green & Associates report:

> Reverse tourism is largely a reflection of the fact that the Maritime Provinces continue to be their own best customers with the Island residents travelling to the mainland and vice versa. (Smith, 1989:34).

Therefore, any gain Prince Edward Island might reap from increased tourism because of a fixed link would be at the expense of tourism operators in Nova Scotia and New Brunswick.

Costs were broken down into capital, operating and maintenance, and estimates of these were provided by PWC. The capital costs of the bridge and rail tunnel options used for the base case were $630 million and $543.5 million, respectively, both measured in 1987 dollars.

FGA set base case (Case 1) assumptions and then performed sensitivity analyses. Nine cases in total were examined. Base case assumptions were:

1. A real discount rate of 10%;

2. Commercial traffic is to increase by 2% annually with or without a fixed link;

3. Passenger-related traffic is to increase by 2.5% annually with or without a fixed link;

4. Induced traffic due to construction of a bridge is 25%, and the same datum due to construction of a tunnel is 15%;

5. The shadow wage rates for permanent jobs and short-term jobs associated with the project are, respectively, 65% and 75% of wages paid;

6. Real ferry-operating costs are to increase by 1% annually;

7. Corporate income taxes generated by either a rail tunnel or a bridge are treated as a benefit; and

8. Time saved by business travellers is valued at the hourly average industrial wage in the Maritimes ($11.38 in 1987), time saved by occupants of trucks is valued at the average truck-driver wage ($16.24), and time saved by non-business travellers is assumed to be half that of business travellers.

FGA then altered the above assumptions (separately) to form an additional eight cases:

Case 2: The real discount rate is 4%;
Case 3: Shadow wages are assumed to be equal to those paid;
Case 4: Capital costs of the rail tunnel and bridge are 20% higher than those used in the base case;
Case 5: Capital costs of the rail tunnel and bridge are 20% lower than those used in the base case;
Case 6: Real ferry operating costs are constant;
Case 7: The value of time saved is 20% higher than in the base case;
Case 8: The value of time saved is 20% lower than in the base case; and
Case 9: Corporate income taxes raised by operation of either a rail tunnel or a bridge are not to be treated as a benefit.

The following table is a summary of FGA's calculations. The least costly option in each case is marked by an asterisk. It reveals that, whereas the rail tunnel option (with or without the Caribou-Wood Islands ferry) is never the least costly option, the other three alternatives could be favoured depending on which set of assumptions one adopts. The existing ferry service option — the status quo — is favoured only in Cases 4, 8 and 9. PWC has signalled its interpretation of the FGA study by entering into an agreement with Strait Crossing Inc. (SCI) of Calgary to proceed with the bridge option with the intention of retaining the Caribou-Wood Islands ferry service.

Net Present Value of Social Costs (in millions of 1987 dollars)

	Status quo	Bridge Only	Bridge & Ferry	Tunnel Only	Tunnel & Ferry
Case 1	528.2	491.8*	492.7	528.4	523.8
Case 2	1,085.0	318.6	267.5*	519.2	454.0
Case 3	647.5	589.8*	606.5	632.0	643.0
Case 4	528.2*	577.8	578.7	602.8	598.2
Case 5	528.2	405.8*	406.7	454.0	449.4
Case 6	489.6	487.7	482.0*	524.3	513.0
Case 7	528.2	455.0	453.7*	508.5	502.2
Case 8	528.2*	528.6	531.7	548.3	545.4
Case 9	528.2*	554.7	555.6	582.3	577.7

Inadequacies of the Analysis

The purpose of this section is to establish sufficient conditions to dismiss the FGA study. This criticism is not meant to be exhaustive, and little effort is devoted to challenging the actual data used. Indeed, we are constrained because FGA do not report all of the data required to redo the analysis. Our principal concern is with a number of elementary, but major, methodological errors. Thus we ignore FGA's failure to investigate shadow-pricing (except of labour), the roles of second-best and peak-load pricing schemes et cetera — the kinds of considerations one would expect in a more rigorous study.

Inclusion of Corporate Income Taxes

For the base case and Cases 3 through 8, FGA include benefits attributed to the raising of corporate taxes — of $62.9 million and $53.9 million for either bridge option and either rail tunnel option, respectively. These benefits are $161.1 million and $138.1 million, respectively, in Case 2. No benefit is attributed to corporate taxes raised in Case 9. Only Case 9 is correct.

There is nothing wrong with including corporate taxes as a benefit to government — part of society — as long as one also counts them as a cost to the bridge or rail tunnel operator, also a member of society. FGA include them as a benefit but not as an off-setting cost, thus biasing the results of the first eight cases in favour of construction.

Most economists would simply ignore changes in corporate tax revenues because they represent a cost to one party and a benefit to another — and thus cancel each other. Indeed, FGA report that they are aware of this problem, and it is the reason they do not include corporate income taxes as a benefit in Case 9. Indeed, in the context of toll revenues, FGA themselves state well the case against including such transfers:

> In keeping with the objectives of the economic analysis, which is to compare society's costs and benefits attributable to each of the transportation alternatives, toll revenues have been excluded from the economic analysis. It should be clearly understood that from society's perspective, the collection of toll revenues does not represent a benefit. Rather it represents a transfer of funds from individual motorists (who are members of society) to the crossing operator (another member of society). No net gain or loss is realized by the transaction. (FGA, 1987:3.2-3.3)

Calculation of Salvage Values

The following is a statement of how FGA calculated salvage values:

> The bridge is expected to have a 100 year life, and thus has a salvage value equal to the ratio of the remaining useful years (65) to the

> expected life (100) times the initial cost ($630 million), which results in a salvage value of $409.5 million in the year 2027. This value must be discounted to 1987. The rail tunnel only option would have a similar salvage value as above, but based on the initial cost of $543.5 million. (FGA, 1987:5.14).

Thus the salvage values of the bridge and rail tunnel, respectively, are calculated to be $9.1 million and $7.8 million in all cases except Case 2 (which is odd, given the assumptions of Cases 4 and 5). They calculate values of $85.3 million and $73.3 million, respectively, in Case 2.

The principal implication of using this method is an absurdity. If a bridge with, for instance, a life of 100 years is worth 65% of its original cost after 35 years of use, then, logically, it must be worth 80% of its original cost after 20 years — and be worth 100% of its original cost at the beginning of the project! To calculate salvage values this way is to assume that the structure is worth what it costs to build, and is thus economically viable. The economic viability of a bridge or rail tunnel is what is to be determined — not assumed.

The proper way to proceed would have been to align the time horizon of the project with that of the assumed economic life of a bridge or tunnel — 105 years: 5 years of construction and 100 years of use. Indeed, this is the methodology that was used in the financial analysis of the fixed link performed by Woods Gordon Management Consultants for PWC. They state:

> The bridge is constructed over the 1988-92 period, and has a life span of 100 years (with regular maintenance). There is no salvage value at the end of the 100 year period. (Woods Gordon, 1987:2).

However, using a 105-year time horizon would call into question the plausibility of FGA's assumptions regarding traffic growth. Peters (1990) reports that 686,500 passenger-related vehicles and 153,000 commercial vehicles made the ferry crossing between Borden and Cape Tormentine in 1989. To examine how plausible FGA's assumptions are, assume that these figures are roughly the relevant traffic data in the year immediately prior to construction of a bridge. Given FGA's base case assumption (d) that induced traffic because of a bridge would be 25%, the number of passenger-related and commercial vehicles, respectively, would increase to 858,125 and 191,250 in the first year of bridge operation. Base case assumptions (b) and (c) are that passenger-related and commercial traffic would increase annually thereafter by 2.5% and 2%, respectively. Therefore, FGA's assumptions would require that, after a further 99 years of bridge use, in the last year of its economic life, approximately 9,890,385 passenger-related vehicles and 1,358,371 commercial vehicles would cross the bridge.

Forecasts of an almost 9-fold increase in commercial traffic and a more than 14-fold increase in passenger vehicle traffic over the life of a bridge both seem quite far-fetched, especially considering that most of the forecasted traffic would occur in July and August.

The question remains, however, that because FGA's analysis does require the calculation of salvage values before the end of the economic life of either option, what should those values be? That is, what would be the (economic) value of a bridge or tunnel in its next best alternative use, after 35 years? One presumes that the only alternative use of a bridge would be scrap, thus giving it a salvage value near zero. The same is true of a tunnel. Obviously, it would be nonsensical to decommission a bridge or tunnel in this manner, but this is an implication of the methodology FGA chose or were instructed to choose.

To understand our criticism of the Case 2 assumption, one must first appreciate why economists "discount" costs and benefits that occur in the future and what this process entails. Consider the problem of receiving $100 today or receiving $200 eight years from today. Even if we assume a zero rate of inflation, these amounts cannot be compared directly because they occur at different times. One way to make them comparable is to express them in values of the same year. By convention, cost-benefit analysts choose the present year. Thus the *present value* of $100 today is simply $100. The present value of $200 eight years from today is the amount that would have to be set aside today and grow at some rate of interest to be worth $200 in 8 years. At an interest (or discount) rate of 10%, $93.30 today would compound to $200 in 8 years. Thus the present value at 10% of $200 to be received 8 years from now is $93.30. The present value of a future sum depends on the discount rate used. The same $200 would have present values of $65.38 and $135.37, respectively, at discount rates of 15% and 5%. The higher the discount rate, the lower the present value of future costs and benefits. Thus the choice of discount rate is critical because it determines how much weight future costs and benefits receive relative to those which occur in the present. Also, the further into the future any amount is realized, the lower its present value. Whereas, at 10%, the present value of $200 to be received 8 years in the future is $93.30, the present value of $200 to be received 40 years from today is only $4.42. For project evaluation, once costs and benefits occurring at different times have been discounted to their present values, benefits may be summed, costs may be summed, and their difference, depending on which is subtracted from what, yields either the present value of net benefits or, as FGA calculate, the present value of net costs.

In Case 2, FGA use a real (net of inflation) discount rate of 4% which exaggerates the relative merit of a fixed link project. To conform with Treasury Board recommendations, a discount rate of 10% should be used, followed by sensitivity analyses using 5% and 15% (Treasury Board,

1976:26). On the other hand, many economists would agree with Burgess' (1981) conclusion that 7% is a more appropriate social discount rate, supported by sensitivity analyses using 4% and 10%. However, without all of Fiander-Good's year-by-year data, it is impossible to perform these calculations. Because low discount rates will tend to favour a fixed link and high ones will favour the status quo, it is necessary to conduct a full reporting of results for a justifiable range of discount rates before an analysis can be completed.

It is assumed in Case 5 that capital costs are 20% lower than in the base case, yet the cost of a bridge has been estimated elsewhere to be $1 billion, in 1989 dollars (Calhoun, 1989:14); that is, over $900 million, in 1987 dollars compared to FGA's base case estimate of $630 million, in 1987 dollars. Using the Case 5 assumption that the capital cost of a bridge is $504 million, in 1987 dollars seems, therefore, to be quite unrealistic.

The Dollar Value of Time Saved

The assumption of Case 7 is that the value of time saved is 20% higher than that assumed for the base case. But, in their base case, FGA use values of time saved which seem quite high. Indeed, because they assume (implicitly) that business travellers and truck drivers are able to make full productive use of every minute of time saved, their base case assumptions in this regard should be treated as maximum values. However, we do not mean to quibble. Our criticism is that it is not necessary to assume dollar values of time saved at all. Given FGA's access to the various roadside surveys they refer to, it would appear that they already possess the data (not published in their report) required to calculate these amounts. Travellers of all types already face a time-distance-money trade-off on the Island when both ferry systems operate in the ice-free season. The methods by which such data can be used to estimate more precisely the value of time saved to all classes of travellers are well known. (A usual starting point for those interested in this literature is Harrison and Quarmby [1969].)

Benefits to Generated (Induced) Traffic

FGA calculate the value of time saved and any increase (or decrease) in vehicle operating costs to be the same for all fixed link travellers regardless of whether they are diverted from the existing ferries or counted as generated traffic. This is incorrect. Whereas this method is appropriate for diverted traffic, "generated" travellers' entire willingness-to-pay for trips via rail tunnel or bridge should be assessed as their benefit. Whether FGA's method overestimates or underestimates the benefit of a fixed link to generated traffic depends on the exact characteristics of generated travellers' demand for this service. The first generated traveller's benefit from a trip via a fixed link alternative is equal to his or her willingness-to-pay for that trip. This amount will exceed the toll charged by whatever extra value this traveller places on a trip by bridge or tunnel versus travelling by ferry. The value of the same trip to the last generated traveller is simply the toll he or she pays — he or she is the *marginal* traveller. The value of a trip by bridge or rail tunnel to other generated traffic lies between these extremes.

Actual Minutes Saved

Beyond the issue of the dollar value of time per hour saved to each class of traveller, there is an important matter concerning how many minutes a bridge would save each type of traveller per trip. FGA state, "Presently at the Borden-Cape Tormentine ferry crossing, an average crossing time, including delays and waiting time, has been estimated to be 100 minutes." (FGA, 1987:5.11). As a bridge trip would take approximately 15 minutes, FGA use a time savings per bridge trip of 85 minutes *for all types of traffic*.

This method, however, overestimates the time savings of commercial traffic. Whereas travellers in automobiles may find their trips delayed, especially in July and August, because they must wait for a ferry with sufficient space, trucks have priority, and are the first vehicles allowed to board. As it would appear that FGA's 100-minute average trip time is based on vehicles of all types, this figure overestimates the minutes saved by commercial traffic and underestimates that saved by passenger-related vehicles. As FGA assume that the dollar value of time saved per hour is much higher for truck drivers than for other travellers, they exaggerate the total dollar value of time saved. To avoid serious overestimation of this benefit, a more appropriate method would be to calculate the average number of minutes saved by each different class of vehicle, convert each to a dollar value separately, then aggregate.

Job Creation — Shadow Wages

Whereas job creation is a political objective of most public investment projects, employment effects are not entered directly in cost-benefit analyses. Instead, they are accounted for indirectly through the use of shadow wage rates.

A shadow wage rate measures the social cost of employing a person on a project, and it need not be equal to the wage actually paid that person. For example, if a project hires a carpenter away from another job, the cost to society of employing that person on the project is the value of production that is no longer produced by him or her in that alternative employment. In this case, the shadow wage is best measured by the gross wage rate actually paid. On the other hand, if a carpenter currently unemployed and looking for work is hired to work on a project, he or she may still be paid the same actual wage as the above-mentioned employed carpenter, but the social cost of hiring that person is less because only less-valued leisure time and not alternative output is forgone.

Therefore, if a project hires only people away from other jobs, actual wages paid would accurately reflect the social cost of hiring them, and the actual wage rate and the shadow wage rate would be one and the same. However, if a project hires a mix of previously employed and unemployed workers, the social cost of hiring them would be less than the wages actually paid, and the shadow wage rate would be less than the actual wage rate paid.

Note that by using shadow wage rates, projects in high-unemployment regions are favoured over those in low-unemployment regions, all other things being equal. This occurs because, although the actual wages paid in both regions may be identical, the use of a lower shadow wage rate in a relatively depressed region would cause social costs as measured by economists to be lower.

FGA's base case assumption (e) is that the shadow wage rates for permanent jobs and short-term jobs associated with the fixed link project are, respectively, 65% and 75% of wages paid. This is to assume that significant numbers of those the project would employ would come from the ranks of the unemployed. In Case 3, they assume shadow wages to be equal to actual wages paid. This is to assume that a fixed link project would only hire people away from other employment — no jobs would be created. One wonders, then, which assumption is more realistic?

FGA report, "... discussions with DRIE [Department of Regional Industrial Expansion] officials in Prince Edward Island indicated that they believe the social opportunity cost of labour is closer to 100 percent rather than 65 and 75 percent." (FGA, 1987:5.11). That is, the project would mostly hire people away from other employment. This is especially true of bridge construction as it would take place mostly during the summer months when employment in fishing and tourism industries is high. It is likely to be less true for tunnel construction, which could proceed year-round.

Three conclusions may be drawn if the advice of DRIE officials is accepted. First, the base case assumptions bias the analysis in favour of construction of a fixed link of either type relative to the status quo. Second, using the same shadow wage rates for both bridge and tunnel options ignores the lower social costs of labour associated with tunnel construction relative to bridge construction. Third, construction of a bridge will not diminish unemployment significantly on Prince Edward Island. If DRIE officials are correct, other government officials and private citizens who applaud bridge construction for its job-creation potential are mistaken.

Ferry Operating Costs

FGA make the base case assumption (f) that real ferry operating costs are to increase by 1% annually over the life of the project. This means that these costs would increase by 1% plus the rate of inflation in nominal terms. In Case 6, they assume that these costs would not increase in real terms and, thus, only by the rate of inflation in nominal terms. The base case assumption thus makes retaining the existing ferry service expensive relative to the bridge and tunnel options and relative to the Case 6 assumption. Certainly, the larger the costs attributed to operating the Borden-Tormentine ferry, the less the status quo is to be seen as the least costly option.

Neither assumption would appear to be consistent with the Report of the Environmental Assessment Panel (EAP). They state: "Between 1979 and 1989, operating costs for the Marine Atlantic vessels crossing the Strait have decreased." (EAP, 1990:9). Thus the base case assumption, especially, would seem to unreasonably bias the analysis against the status quo.

Other Options?

Another base case assumption is that a bridge would generate 25% more traffic than the existing ferry service. This figure for a rail tunnel is 15%. This difference is, presumably, attributable to FGA's calculations that a bridge would save 85 minutes of travelling time per trip on the Borden-Tormentine run, whereas a rail tunnel would save only 55 minutes because part of the trip by train would involve time spent loading and unloading vehicles. Aside from the matter of how accurate these figures are, one option FGA do not consider — they were not asked to — is a highway tunnel. If they had, and had they made the same time saving and generated traffic assumptions for it as they did for a bridge, a highway tunnel would have dominated the bridge option in all cases, using their methodology (unless there is some extra, significant cost associated with a highway tunnel and not with a rail tunnel).

Moreover there are other reasons beyond just lower capital, maintenance and operating costs to prefer a highway tunnel to a bridge. (Although FGA attribute higher operating costs to a rail tunnel than to a bridge, it seems unlikely that it would cost more to maintain a highway tunnel than a bridge.) First, it would be safer for travellers and far less prone to closure in bad weather. Second, a major environmental danger of a bridge stems from the possibility that it would impede ice break-up in the Northumberland Strait each spring. This, in turn, has impacts on fishing and agricultural industries.

Another bridge-related problem has to do with its impact on silting in the waterway. Marine populations are, obviously, especially sensitive to this aspect of the marine environment. Whether fears of these impacts are exaggerated or not, these same fears do not hold for a tunnel, which is the choice of environmentalists, both in Canada and abroad, and some engineers. (See Land [1989], Staff [1990] and Dyck [1991] on this matter.) Indeed, even the Generic Initial Environmental Evaluation (GIEE) conducted in compliance with federal environmental assessment regulations awarded the bridge a much higher risk-score than the tunnel: 547,443 to 139,102 (Perchanok, 1988:14). Although the costs of uncertain environmental damage and risk to life and limb may be difficult to measure, analysts can appeal to substantial literature on the theory and measurement of both. There is no justification for ignoring them in a study meant to assess all social costs and benefits. A more thorough approach would seem especially prudent given that in its August 1990 Report, the Environmental Assessment Panel recommended that bridge project not proceed.

This is not to say, however, that a highway tunnel would be preferred to the status quo. A more complete study would have considered as many technically feasible options as possible, including, as Peters (1991) details, the use of faster vessels.

Conclusion

There are, then, sufficient grounds to dismiss both the Fiander-Good study and Public Works Canada's conclusion that construction of a bridge between New Brunswick and Prince Edward Island is justified on economic grounds. Although FGA's conclusion is that society should be indifferent between a bridge and existing ferry options, methodological errors in their analysis tend unduly to favour construction of a fixed link, especially a bridge. It would not be surprising, however, if these errors were the result of FGA following PWC's directive to produce a study of the "Stanford and the DRIE-TC-PWC" type.

Regardless of origin, these mistakes and omissions are major. Indeed, when FGA's errors concerning corporate income taxes and salvage values are corrected, and when the extreme assumptions of cases 2, 5 and 7 are dismissed, the status quo dominates in all scenarios examined. Other corrections simply reinforce this conclusion: that is, construction of a fixed link would impose a net loss on society. Seen from this perspective, the debate concerning the magnitude of any ensuing environmental damage would seem moot: Any damage at all would simply magnify this inevitable social loss.

It is important that decision-makers understand that the economic viability of a fixed link has not been established. At the very least, Public Works Canada would be well-advised to review their decision to proceed by commissioning a more complete and methodologically sound cost-benefit analysis of this project.

References

Burgess, D.F., "The Social Discount Rate for Canada: Theory and Evidence," *Canadian Public Policy VII*, 1981, pp. 383-394.

Calhoun, S., "When PEI joins the mainland: How much will a fixed link change Island life?" *Canadian Geographic 109(2)*, April-May 1989, pp. 12-21.

Dyck, H., "Despite calls for tunnel option, MacKay backs idea of bridge to PEI," *The* Chronicle-Herald, 2 January 1991, p. C12.

FEARO (Federal Environmental Assessment Review Office), "Report of the Environmental Assessment Panel," 1990. (Ottawa).

Fiander-Good Associates Ltd., *Draft Final Report: Economic Feasibility Assessment for the Northumberland Strait Crossing*, 1987. (Fredericton).

Harrison, A.J. and D.A. Quarmby, "The value of time in transport planning," in *Theoretical and Practical Research on an Estimation of Time-Saving, 1969. European Conference of Ministers of Transport, Report of the Sixth Round Table (Paris).*

Land, T., "Greens may dictate Baltic fixed link," *The Chronicle-Herald*, 7 August 1989, p. A7.

Perchanok, N., *Backgrounder: The PEI Fixed Link Project*, 1988.(Ottawa: Science and Technology Division of the Library of Parliament).

Peters, T., "Island ferry service still in question," *The Chronicle-Herald*, 2 June 1990.

Peters, T., "High speed vessels seen as replacements for existing ferries," *The Chronicle-Herald, 2 March 1991, p. A16.*

Smith Green & Associates Inc., *Bridge Concept Assessment Supplement: Effects on Tourism and Associated Development Strategies — Support Document for Question J*, 1989. (Ottawa: Public Works Canada).

Staff, "Consultant forecasts PEI tunnel link," *The Chronicle-Herald*, 24 December 1990, p. A2.

Townley, Peter G.C., "The weakest link: The economic viability of a Northumberland Strait crossing," 1992. *Policy Options XIII(6)*: pp. 15-18.

Treasury Board, *Benefit-Cost Analysis Guide*, 1976. (Ottawa).

Woods Gordon Management Consultants, *Financial Analysis of the Northumberland Strait Crossing Project*, 1987. (Toronto).

Financing a Bridge: The SCI Plan

Donald Deacon

In this article I am concerned with the short term plan to finance the proposed bridge. I will offer an analysis of the proposed Strait Crossing Inc. (SCI) financing, outlining its probable effects and immediate shortcomings.

The SCI Financing Plan

Before construction on the proposed bridge begins, the federal government requires SCI to place in the hands of a financial trustee all the funds which it believes will be necessary to complete the bridge. SCI proposes to raise the estimated $840 million required to build the bridge, in the following manner:

$600 million through debt securities paying interest at 4.75%, interest and principal payments on which will be secured by the indexed annual federal government subsidy of $42 million for 35 years. These payments are to commence in October, 1997, the scheduled completion date of the bridge.

$150 million through the sale of Partnership Units, the owners of which will be entitled to share in the toll revenues of the bridge when it is completed. Prior to that time, the owners of the Partnership Units would be able to write off portions of their investment against their other income.

$90 million from interest earned on the unspent portions of the funds raised through the above sale of debt securities and Partnership Units. This estimate of interest income is based on construction costs being spread over a period of 4 ½ years.

SCI's "Real Interest" Bond Issue

"Real Interest" bonds are bonds upon which the interest and principal payments are adjusted annually by the Consumer Price Index (CPI) so that the purchasing power of interest and principal is always maintained. The Government of Canada sold its first "Real Interest" bonds in December, 1991 when it successfully borrowed $700 million at 4.25%, due in 2021. It borrowed an additional $500 million in December 1992. However, by that time the price of the first bonds had declined about 8% in value and the second issue was sold at 92.15% of its face value to yield 4.75%. Although a "normal" Government of Canada long-term bond currently yields about 8.5%, inflation would have reduced its purchasing power by about 1.5% in 1992, producing a "real" yield of 7% (8.5% less 1.5%). The difference in the "real" yield between the "normal" bond (7%) and a "Real Interest" bond (4.75%) is thus 2.25%. The principal amount owed is also increased by the CPI, further reducing the difference in ultimate return after inflation. Some managers of non-taxable funds such as pension funds are prepared to give up the higher yield for reliable protection against inflation. However, the difference in "real" yield and the fact that the inflation increase is subject to taxation is the reason "Real Interest" bonds are not attractive to all institutional investors, particularly those whose income is subject to income tax.

How can SCI expect to borrow $600 million at an interest rate of 4.75% or less to build a bridge under such risky conditions? It can, for two reasons. The first is that SCI's annual payments from the federal government will be indexed to inflation. SCI will therefore be able to increase the interest and principal payments each year by the amount of inflation. The institutions which purchase SCI debt securities can thus be assured that the purchasing power of the interest and principal will remain constant throughout the 35 years. In addition, the Act Respecting the Northumberland Strait Crossing (Bill C-110), will effectively provide a Government of Canada guarantee. SCI's prospective buyers are insisting that the indexed subsidy payments be paid into a trust fund because they are not prepared to accept any risk of failure to perform by the SCI consortium. The government is, in this way, arranging to have Canadian taxpayers guarantee over 85% of the financing of this so-called "free enterprise" mega-project and bear any costs and liabilities associated with the bridge which the SCI consortium is unable or unwilling to cover.

Interest payments on the SCI bonds are not to begin until the scheduled date of completion. At that time, the government is committed to begin paying the annual subsidy of $42 million into the trust fund, indexed each year, regardless of whether the bridge is completed, is paid for, or is operated and maintained in a satisfactory manner. No dispute which might arise between the government and SCI will be allowed to interrupt payments of bond interest and principal to SCI's bondholders. Interest during the 4 ½ year construction period will be compounded and added to the original $600 million principal. Thus, when interest payments finally do begin after 4 ½ years, the outstanding debt will exceed $800 million, and annual interest payments will be around $38 million.

SCI's Partnership Units

SCI's Partnership Units will be a form of ownership that can provide tax write-offs prior to completion of the bridge. Investors would eventually participate in the bridge's toll revenues, at which time any profits received would be subject to income tax. Partership investments have been a common form of financing oil and gas exploration and similar risk developments. They are of interest to investors who seek tax deferrals and are prepared to take a risk. It is an interesting way for the SCI consortium to raise the required shareholder funds without having to invest more than it has already risked to date in developing the project and endeavouring to win the contract.

The net funds which SCI could expect to realize from the sale of $150 million worth of Partnership Units might be as much as $140 million after commission, legal and other expenses.

Interest on Unspent Funds

To determine the validity of SCI's estimate of $90 million as the amount it will earn on the unspent portion of the $740 million, an estimate of costs in the SCI construction schedule is required. Unfortunately, the relevant information is unavailable at the time of writing.

Toll Revenues

The federal government's proposed contract with SCI provides that annual toll increases must be limited to 75% of inflation, except under two circumstances. The first of these mitigating circumstances would occur if the government requires additional insurance coverage not already specified in the agreement — the cost of the additional coverage can be recovered by increasing toll rates. The second circumstance would occur if the traffic count falls below the traffic level in the year before the bridge opens. In this case, the tolls can be increased in the following year sufficiently to recoup any loss that was experienced.

Had the federal government been indexing ferry subsidies during the last 10 years when inflation totalled 60%, ferry tolls could have been lower in 1993 than they were in 1982. Because the operating subsidy was not increased, toll revenues had to be increased by 125% in order to cover the full impact of inflation during the period. This has proved very beneficial for the financing of the fixed link because the latter would not have been feasible if toll revenues were only 40% of present levels.

Summary of SCI's Financing Plan

From the above information, one can determine what SCI estimates the cost of construction of the 13 kilometre bridge and its 14 kilometres of approach roads to be. Their estimate appears to be as follows:

Net cash raised from sale of Partnership Units:	$140 million
Net cash from sale of 4.75% bonds:	$600 million
Approximate cash received from above:	$740 million
SCI's estimate of revenue from unspent funds:	$90 million
Total cash available for total project:	$830 million
minus SCI's Contingency Fund:	$100 million
minus SCI Fisherman Compensation Fund:	$10 million
Net cash for construction and overhead:	$720 million

So, $720 million appears to be SCI's estimate of the actual cash required to build the $840 million bridge. As long as cost overruns of this mega-project, which will be constructed under the challenging conditions of the Northumberland Strait, do not exceed 35%, SCI and its investors (as well as the federal government and Canadian taxpayers) should be safe. However, the cost overruns and completion delays experienced in some other major and innovative fixed link projects have been substantial: the English Channel tunnel was 90% over budget; the Penang Bridge was 70% over budget; and the Jamestown-Verrazzano bridge, Providence, R.I., was 132% over budget and took three years longer than estimated to complete!

Because the trust fund of the SCI bond issue will make it effectively an obligation of the Canadian government, it would not be affected by an SCI default. Construction cost overruns, failure to adequately maintain the structure and other operational concerns will be matters solely between the federal government and SCI.

The true cost of the bridge project is now well beyond the annual indexed payment to be made to the SCI bondholder's trust fund. Justice Barbara Reed of the Federal Court of Canada estimated that Public Works Canada (PWC) has spent over $20 million since 1985 on studies and procedures. In addition, the government now proposes to give $20 million to Borden and Cape Tormentine to help offset their losses arising from the shutdown of Marine Atlantic, and plans to give $40 million for extra highway construction to N.B. and PEI.

Both the federal and the PEI goverments will bear responsibility for the consequences of the loss of almost $40 million of annual revenue now generated by Marine Atlantic through wages, benefits, supplies and services for the ferry service. Damage to the annual $75 million Northumberland Strait fishery beyond the $10 million to be assumed by SCI, will also be a major responsibility.

But the most serious potential liability is that of Transport Canada in the event of an extended disruption of bridge traffic caused by a ship collision or a serious accident during the 5 winter months when there is no alternative access to the mainland. Paul Giannelia, SCI's President, told listeners at one of the public meetings in PEI that replacing one section of a damaged bridge would take 18 months. Public Works Canada's response to this, so far, has been the statement that there will be plenty of time to consider alternative crossing options during the 5 years when the bridge is being built.

In short, it is clear that the plan to finance the construction of the SCI bridge leaves too many unanswered questions and is seriously flawed.

Decommissioning the Fixed Crossing: The Forgotten Cost

Joseph H. O'Grady

The proposed construction of a bridge across the Northumberland Strait from New Brunswick to Prince Edward Island has been hailed as one of the greatest engineering challenges in Canadian history. At the same time, investors assert that it is technically feasible and their financial forecasts predict the economic viability of the project. But before contracts are signed and blueprints become reality, it is important to offer a reminder about the temporary nature of all assets and the need to prepare for their obsolescence. It is not "doomsday accounting" to recognize that, in the long run, *every* Strait crossing option will be temporary. Ferries will depreciate to rusting hulks; tunnels will collapse; and bridges will experience enough structural deterioration to make them unfit for further service. A critical question becomes, "Then what?"

Experience confirms that an obsolete ferry can retain some salvage value, if only for scrap metal. But will a *bridge* still have a salvage value when its functional life is over? Or, might it accrue a *negative* salvage value? Could its demise create an economic liability for someone at some time? Is this liability a "forgotten cost" in the feasibility equations? Upon whose shoulders should this cost fall? Can the cost be anticipated to ensure fair allocation? These are some of the questions addressed in this paper.

Obsolescence and Externalities

All of us have experienced the phenomenon of product obsolescence. We can all recount incidents when a consumer good or a business asset wore out, broke down, fell from style or lost appeal because of new alternatives. What everyone may not understand is that there are at least two sources of product obsolescence: functional and performance. An existing product becomes *functionally* obsolete upon the arrival of a newer, technically superior substitute. *Performance* obsolescence is experienced, sooner or later, when a product simply wears out.

Businesses and individuals react differently to the inevitability of product obsolescence. Few consumers plan for it; we simply accept the fact that nothing lasts forever. When an item expires, we dispose of it and replace it. Businesses generally take a more strategic stance regarding asset obsolescence. Through the process of depreciation, an allowance is made for the decrease in value of an asset over time. The cost of the asset (less its salvage value) is gradually written off over its useful life until the outdated asset is removed from the books of the organization. In either case, the asset is transferred out of the hands of the owner and eventually its burden falls elsewhere. Recycling facilities, salvage yards, landfills and incinerators represent some options for the disposal of obsolete products.

The production, consumption and disposition of "temporary satisfiers" is an acceptable practise for most of us in North American society. And no economic argument can be made against this right to consumer sovereignty, provided the buyers know products have limited life spans, and provided they absorb fully all the costs of depreciation and obsolescence associated with their purchases. But if the costs are not fully internalized, and if they begin to overflow onto other members of society, then a different situation exists.

Economic externalities, or "spillovers," describe situations where costs and benefits of activities are not fully included in their market prices. Externalities inflict economic harm without offering compensation or, occasionally, confer gains without requiring payment. The classic example of an externality is pollution. When factories dump their wastes on public land or into water or air, they transfer an uncompensated burden onto other members of society. The dumpers are not fully absorbing all the costs of their production, consumption and disposition decisions. Most economists agree that when markets fail due to externalities, a *prima facie* case exists for government intervention to assist those markets to reach greater efficiency.

The concepts of obsolescence and externalities come together in situations where businesses and consumers fail to prepare for the responsible disposal of their obsolete assets. Since the price tag at purchase generally does not include the cost of disposing of an asset when its physical or economic life is over, this cost may be shifted onto the shoulders of society. Obvious examples of this include deserted buildings in inner cities, eroded timber lands, abandoned strip mines, toxic waste sites, overloaded landfills and the associated pollution of our natural resources.

The Obsolescence of Strait Crossing Options

A little reflection should leave no doubt in anyone's mind that *all* Northumberland Strait crossing options are temporary. Older residents of the Island have seen several passenger ferries fully depreciate due to time and the elements faced in crossing the Strait. (The most recent example of this, of course, is that of the famous *Abegweit I*.) There is also evidence that bridges become obsolete. Trade journals, such as *Engineering News-Record*, often report instances of bridge deterioration due to faulty components, poor design, structural deficiencies, construction shortcuts, salt damage, erosion and vehicle accidents. Responding to several bridge collapses in the late 1980s, the United States Federal Highway Commission reported that 238,000 bridges in America — nearly 50% of the total number — were rated as "deficient" (Eldred 1989, 67).

Clearly, the transportation infrastructure is subject to performance obsolescence. Yet, the two major studies evaluating the fixed link options and impacts offered no special mention of this. The closest these reports come to acknowledging obsolescence are references to a 100-year life, for the convenience of discounting costs and benefits (Fiander-Good 1987, 5.14; Woods Gordon 1987, 2). Needless to say, there are many skeptics who scoff at the intimation that any man-made structure would last as long as 100 years when exposed to the adverse weather conditions of the Northumberland Strait. It is worth noting, too, that functional obsolescence could affect a bridge much earlier than performance obsolescence. The creative abilities of engineers will not peak when the last rivet is driven and the last ton of concrete is poured. Engineers will continue to invent, refine and improve. No one can predict what new energy sources, building materials, engineering processes or transportation alternatives will be available as we enter the twenty-first century.

Clearly, engineers cannot guarantee that the crossing option chosen in the 1990s will still be the best choice in 25 or 50 or 100 years. There is a chance that technological advances will allow a more efficient alternative to render any 1990s option functionally obsolete. Yet, nowhere are there plans that acknowledge or prepare for this contingency.

The Concept of Negative Net Salvage Value

The certainty that every Strait crossing option will experience functional and/or performance obsolescence is not in itself a valid reason to halt construction of a bridge. However, an understanding of the issue should encourage us to consider what values the various crossing alternatives will retain after they have fallen into obsolescence, and how those values could affect society. Specifically, we ought to examine the concept of *net salvage value.*

Net salvage value is derived from two factors: 1) the cost of physically removing an obsolete asset from service; and 2) the value of that asset to a subsequent owner. Net salvage value is the difference between these two components. Normally, accountants anticipate that a positive net salvage value will result from an asset. Either removal costs are low enough, or salvage worth is high enough, to expect a positive return. Usually this monetary incentive is sufficient to encourage disposing of the asset in a responsible manner, rather than simply abandoning it.

Certainly this is the process associated with the disposal of obsolete passenger ferries. Occasionally these ships find limited service in some less demanding nautical capacity. (The original *Abegweit* is a good case in point: She has found new life as a floating restaurant and club facility in Chicago.) More often, though, obsolete vessels are sold to marine salvage yards where they are stripped of valuable hardware, cut up for scrap metal and recycled.

It is significant that even in a fully depreciated state, a passenger ferry can still retain a positive net salvage value. There is comfort in this fact for anyone concerned about environmental issues. Recently, however, economists and accountants have confronted the reality of obsolete assets with *negative* net salvage values. Although the concept may be unfamiliar to people outside the public utilities industry, there is growing recognition that greater attention needs to be paid to it. For instance, utilities experts warn that the high costs of decommissioning (shutting down) obsolete nuclear reactors will result in significant negative salvage values for them. The costs of safe removal and disposal of contaminated nuclear assets will greatly exceed their value to subsequent owners.

It is not a major leap of reasoning to suggest that an obsolete 13 kilometre bridge also could result in negative salvage value. The future cost of its removal and disposal could well exceed its salvage value. Realistically, it is improbable that the rusted steel and weathered concrete would have any value to a subsequent user. Although recorded as an asset in the books, an outmoded bridge of this size and type could be classed more accurately as a liability to its owners and, ultimately, to society.

Economic Implications of Negative Net Salvage Value

One of the two factors that determines net salvage value is future removal cost, and the major uncertainty in predicting removal cost is inflation. Over a long time period, inflation rates become critical. (A 5% annual rate of inflation would nearly double removal costs every 15 years.) If inflation for the projected life of the bridge is factored, the potential cost of removing it in 50 or 100 years becomes an important issue. When this concern is combined with the unlikely prospect of finding a subsequent user to purchase the salvaged materials, we have the two ingredients needed for a classic case of *negative* net salvage value.

Each of the leading consultants' reports gives only brief reference to the critical issue of salvage value. The Fiander-Good analysis, based upon a 40-year time frame (1987 to 2027), suggests only that, by the year 2027, the 35-year-old bridge will have retained a salvage value of $409.5 million (Fiander-Good 1987, 5.14). But this estimate offers little relevant information to the analysis. The $409.5 million figure reflects only the arithmetical fact that, after 35 years of straight line depreciation, a $630 million bridge has $409.5 million remaining to be "written off" in the books. Following this format, it will have a book salvage value of "$0" at the end of 100 years.

It should be obvious even to non-accountants that this depreciation procedure does not offer a fair assessment of the actual market worth of an obsolete bridge. This *book* salvage value in no way reflects the two key components of net salvage value identified earlier: the cost of removing the obsolete asset from service and the value of it to subsequent owners after it is removed. Even if we accept the incorrect "zero worth" scenario, a serious concern persists: If an asset has no residual salvage value, then where is the monetary incentive to remove it and dispose of it?

The Woods Gordon report also downplays the role of salvage value to the fixed link analysis. The report states: "We ignored the possibility of salvage values for all assets in 2092, since these would not have significant implications when discounted back to 1987" (Woods Gordon 1987, 2). This point is valid only if they are considering the possibility of a minor, *positive* salvage value. If a major, *negative* salvage value accrues 100 years hence, it could have significant implications, even when discounted 100 years back to the present. This possibility has not even been considered in the financial analysis.

The people of Prince Edward Island and New Brunswick may see the construction of a project that will become a financial liability for someone at some point in the future. Still unaddressed is the question of who will accept the financial responsibility for the structure's eventual removal. This responsibility is clouded by the terms of the construction agreement which stipulate that, after a period of 35 years, ownership of the fixed link assets will transfer to the Government of Canada. The builders and initial operators of the link will walk away from their project just as it enters the critical "middle-age" years of its life span. They, and the first generation of link users, will avoid financial responsibility for dismantling the project from which they will have derived significant revenues. Instead, this burden is destined to fall on a future generation of Canadian taxpayers. Arguably, this amounts to the granting of a *de facto* subsidy to the bridge builders, equal in dollar value to the future costs of decommissioning the project.

Fortunately, there are procedures to ensure that the builders, operators and users of a fixed link will absorb all its costs, including construction, demolition and disposal. The potential externality can be internalized and targeted to those users who ought to bear its full burden.

Options for Funding Bridge Decommissioning

Planning for the abandonment of the fixed link will be a complex procedure, laden with uncertainties. Someone will have to provide an indefinite amount of money at some unpredictable time in the future, to address an unprecedented challenge: the safe demolition, removal and disposal of the 13 kilometre bridge. This planning must be done without reliable estimates of the costs, technologies or economic circumstances which could prevail at that future date. Considering this climate of uncertainty, and the consequences of error, there is room for only the most conservative financial planning. Several funding options exist, but just one offers the security and stability that this decommissioning demands. For the purposes of discussion, however, a range of options is summarized below.

Option #1: treating removal costs as an expense. This option is hardly worth considering, but its futility ought to be demonstrated. "Expensing" presumes that the full cost of removing the bridge could be generated from normal revenues collected in the year of removal. Demolition expense would be treated as would an electricity bill or workers' salaries.

But, it is unrealistic to presume that any significant amount of money might be generated from current revenues. Even if funds could be raised during the bridge's final accounting cycle, to do so would be grossly unfair. The last generation of users would shoulder the full burden of dismantling the bridge from which they received relatively little service. Prior generations of users, particularly the one which endorsed the bridge's construction, would escape scot-free. Clearly, expensing removal costs is the antithesis of good contingency planning. Ironically, in the absence of an explicit decommissioning plan, expensing will become the *modus operandi* by default.

Option #2: bond funding. This method would also postpone acquisition of demolition money until the last moment. In theory, the final owners of the fixed link (apparently, Public Works Canada) would float a bond to finance removal costs when demolition becomes necessary. The debt would be repaid from the cash flow of other revenues (taxes) generated by the bridge owner.

But borrowing capital for bridge demolition could be an expensive proposition. Investors may demand a high return (interest rate) to cover the perceived risks of such a project. If large cost overruns are anticipated, investors may balk altogether. Bond-funded removal of a bridge offers security under only one extreme set of circumstances: if planners can accurately forecast the economic climate, the Government of Canada's debt capacity, the market's ability to sustain debt, inflation rates and interest rates at various future dates. Unfortunately, these factors cannot be predicted with any degree of accuracy.

The other criticism of this option is the same one associated with Option #1: It is inherently unfair to shift the decommissioning burden away from current owners and users and onto future taxpayers.

Option #3: surety bonding or insurance. In principle, the bridge owners would contract with a surety bonding company, such as Lloyd's of London. In the event of the owner's failure to perform specified acts (in this case the demolition, removal and disposal of a bridge), then the bonding company would fulfill the obligation. For this privilege, the bridge owner would pay a premium (probably in annual installments), that would represent the regularly adjusted default risk involved. The fees would become a normal expense of doing business and they would be reflected in the rate structure for bridge users.

Although it looks appealing, this option is not a panacea. Normally, surety bonding is designed to play a backup role, not to serve as the primary vehicle for funding removal costs. No bonding agency could provide this type of high risk coverage at affordable rates unless a responsible plan was already in place to fund removal costs. Even with such a plan, the annual premium needed to cover the cost of guaranteeing such a risky venture would be significant.

Option #4: dedicated demolition fund. Making periodic deposits into a demolition fund is another way to accumulate the needed money. Bridge users would be charged an additional fee with each crossing that would be credited to a demolition account. The size of the fee would depend on life expectancy of the bridge, projections of demolition costs, net salvage value and number of bridge users over its lifetime. Ideally, this account would be segregated and managed independently to prevent diversion of funds into general revenues. Periodic re-estimations of all the variables (and corresponding adjustments to user rates) would be required to reduce risks and preserve the integrity of the fund.

Dedicated reserves are often recommended by public utilities experts as a responsible way to fund the decommissioning of nuclear power plants. The same rationale would apply to the dismantling of a bridge with an anticipated negative salvage value. But even a dedicated fund would be inadequate if the bridge succumbed to obsolescence earlier than expected. This circumstance would render inaccurate all fee calculations and result in a funding shortfall. An insurance policy (surety bonding) to protect against premature obsolescence would be a necessary hedge to protect the dedicated fund in the event of premature closing of the project.

Option #5: pre-established funding account. One way to promise that demolition funds will be available is to require the link builder to deposit all the needed money into a segregated, independently administered account *before* the bridge enters service. Interest income would increase the yearly balance of the account, helping to offset the problem of inflation. This option would still require an accurate estimate of the date of obsolescence so the initial fund amount could be determined. But, if removal cost estimates are correct, and if additional deposits are made by subsequent operators when cost estimates are periodically revised, sufficient funds will be available when they are needed. Surety bonding could offer the extra measure of confidence needed to ensure the availability of this demolition reserve.

Obviously, this would be the most expensive option for the bridge owners and its cost would be transferred to bridge users through higher crossing fees. There is also a tinge of unfairness associated with this funding method because it would require bridge users to pay up-front for an expense which may not be incurred for many decades into the future. But, arguably, it is better for current users to bear a small inequity than for future users (or taxpayers) to shoulder a larger one.

This brief overview of funding options is not exhaustive; hybrids and variations are possible. And there are many complicating issues, such as current accounting and tax considerations, that have to be factored into each option. However, it should be apparent that guaranteeing the funds for removal of a fixed link may not be easy or cheap. Just like ordinary insurance policies, the funding methods which offer the greatest protection and solvency will be the most expensive to implement.

Conclusions and Recommendations

As the Canadian government prepares to grant permission for construction of a Northumberland Strait crossing, a number of vital concerns remain unaddressed. Among them are these:

1. No commitment has been offered to Prince Edward Islanders that they will not be affronted aesthetically by a massive bridge superstructure long after it has been forsaken for economic or physical reasons. Just as abandoned strip mines, quarries and clear-cut forests mar parts of our Canadian landscape, the remains of such a bridge could continue to deface our Island seascape long after its viability has passed. Commuters between Charlottetown and points east receive a daily reminder of how easily this can happen. For nearly 30 years, the abandoned footings of the old Hillsborough Bridge have been a blight to the eye as, stone by stone, they erode and collapse into the harbour waters. This same scenario, on a far greater scale, could easily unfold in the Northumberland Strait.

2. No promise has been made that the Northumberland Strait will be returned to its original state when the bridge has been rendered obsolete. We have the moral responsibility to transfer the waters of the Northumberland Strait to succeeding generations without environmental liabilities or aesthetic liens attached.

3. No plan is in place to assure future generations of Canadians that they will not bear the economic liability for construction decisions made (and demolition plans ignored) in the 1990s. Providing for the bridge's demolition costs will be a challenging task, but it is not one that can be ignored. Astute financial management may allow the builder to write off the bridge in the books and emerge financially secure. But these manipulations will not erase the mega-project from the Northumberland Strait. Economists chastise us often enough for passing on our provincial and national debts to future generations, so that we may live beyond our means today. The same admonition should accompany any plan to shuffle the burden of dismantling a fixed link onto future generations of Canadians.

Several recommendations flow from these conclusions. First, the issues of obsolescence and removal costs need to be resolved. Each firm bidding for the privilege of building a fixed link should identify its perceived responsibility for the project's removal. Disavowed liability should be a major consideration in the granting decision. If bidders acknowledge responsibility, we should be informed of specific plans to finance the removal costs.

A second suggestion is that the two feasibility studies that paved the way for the project (Fiander-Good and Woods Gordon) be reworked to reflect the decommissioning concerns. As they stand, they are incomplete because they do not include consideration of all the costs. The discounted benefits of a bridge will decrease due to the inclusion of more realistic net salvage values; the discounted societal costs will increase due to higher capital costs associated with a removal fund; the combined result will be an increased net present value of societal cost.

A third recommendation is that the Province of Prince Edward Island consider drafting legislation that would outlaw the abandonment of any major asset. It would require that builders of projects infringing upon public natural resources accept the responsibility for their eventual removal. The demonstrated willingness and ability to do so would be a mandatory prerequisite to construction. This would encourage engineers and architects to consider demolition during the design stage and improve the efficiency of the process.

Certainly, it is tempting to push these worrisome thoughts aside and focus instead on the savings in travel time, the convenience, the increased tourist traffic and the temporary job creation promised by a fixed link. But if we fail to consider the full economic and social costs of the project, if we ignore the certainty of its obsolescence, if we lull ourselves into believing the problems will disappear on their own, then we are only fooling ourselves. The debts will be incurred, even if the bills are deferred. The only questions unanswered are "*Who* will end up paying them ... and *when*?"

References

Eldred, William, "Bridge Safety: Who's Looking Out For Us?" *American City & County* 104 (Sept. 1989): 67-73.

Fiander-Good Associates Limited, *Economic Feasibility Assessment for the Northumberland Strait Crossing*, 1987.

Woods Gordon Management Consultants, *Financial Analysis of the Northumberland Strait Crossing Project*, 1987.

The Economic Impact of the Withdrawal of Marine Atlantic Ferry Service from Prince Edward Island

Prepared for the Canadian Labour Congress by the Cooper Institute

The Northumberland Strait, even where it shrinks to its slightest at Abegweit Narrows, remains 13 irrefutable kilometres of blue-grey pack ice between Borden, Prince Edward Island and Cape Tormentine, New Brunswick. It is an undeniable constant, a geographic certainty of Prince Edward Island. Any thoughts about land-based transportation which connect Prince Edward Island with the mainland must pay heed to this element. Whether the infrastructure is a bridge, a ferry or a road or rail tunnel, the flow of traffic through these 13 kilometres will always be required to salute the essence of islandness — that an island is surrounded by water.

Even in the absence of toll booths and adverse weather conditions, all land-based traffic will take longer to travel through those 13 kilometres than just about any other 13 kilometres in the country. Special restrictions will be in place for the transport of dangerous goods through these kilometres; lower speed limits will be a constant reality for bridge and road tunnel travel; weather conditions will affect ferry and bridge travel; and loading/unloading will slow rail tunnel and ferry travel. Delays and slowdowns are unavoidable.

Because of the importance of a transportation link to Prince Edward Island, proposed changes in transportation infrastructure must be closely examined. The benefits to the Island economy of the proposed bridge project must be shown to be greater than the benefits of the ongoing presence of Marine Atlantic. We cannot afford shortsighted vision — because this is a project with a 100 year life, we must have a vision for the future and what the next millenium may bring in transportation innovation.

Today, proponents of change to Prince Edward Island's Northumberland Strait transportation infrastructure must weigh the benefits of a road-only bridge against those of both the ongoing economic presence of Marine Atlantic on the Island economy and of the versatility this service gives to current and future modes of transportation.

The subject of this report focuses specifically on the economic impact of the withdrawal of Marine Atlantic services from Prince Edward Island. It also aims a spotlight at several reports which discuss the economic impacts of the construction/operation of a proposed bridge on the PEI economy. Finally, it looks at future trends in road transportation and their implications for Prince Edward Island.

The Economic Impact of Marine Atlantic

Marine Atlantic is a Crown Corporation which operates 6 ferry services in Atlantic Canada as well as a coastal service in Newfoundland and Labrador. The Prince Edward Island service is Marine Atlantic's busiest crossing and in 1991 it carried 1.7 million passengers across the Northumberland Strait.1

Marine Atlantic is one of Prince Edward Island's largest employers. Its unionized work force is relatively well paid with secure employment and good benefits. Construction of a bridge between New Brunswick and Prince Edward Island will mean the closure of Marine Atlantic's Northumberland Strait ferry service. The loss associated with this proposal will cause serious problems to the Island's social and economic structure.

Economic inputs to the Island's economy from the ferry service have been divided into four components: employees' wages, purchases of supplies and services for Island businesses, spin-off effects in the Prince Edward Island economy, and employment and income from cafeterias and gift shops located both within the ferries and at each ferry terminal.

Wages of Marine Atlantic Employees

The Marine Atlantic ferry service between Borden and Cape Tormentine employs a total of 651 persons — 402 are permanent positions and 249 are seasonal. Their wages represent an input of $19.5 million and an additional $4.8 million in benefits to the regional economy. Ninety per cent (90%) or 363 of Marine Atlantic's permanent employees, and 77% or 191 of their seasonal employees, live in Prince Edward Island. Seasonal employment represents 35 to 50 summertime jobs for students and 140 jobs averaging 6 months each. These latter, longer-term jobs are often supplemented by unemployment insurance benefits and constitute full-time incomes. Wages alone amount to $16.8 million of input into the Prince Edward Island economy each year. Each of Prince Edward Island's three counties enjoy a portion of this employment. The location of the ferry terminal in Borden, Prince County, accounts for the concentration of the greatest proportion of employees (74%) in that county.

Closure of the Marine Atlantic ferry service would have a greater effect on Prince County than on either Queens or Kings counties. Already, Prince County experiences the highest level of unemployment of all counties in Prince Edward Island. Employment statistics obtained from Employment and Immigration Canada indicate that the seasonally-adjusted unemployment level in Prince County in May, 1992 was 19.1%.

Prince Edward Island's small rural communities are especially vulnerable to declining employment opportunities. Out-migration of unemployed ferry workers to urban centres in Prince Edward Island and other parts of Canada are among the predictable social and economic casualties of the termination of Marine Atlantic as an employer on Prince Edward Island.

Loss of secure jobs such as those at Marine Atlantic cannot be measured only in dollars and cents. They also result in deterioration of the rural fabric of the Atlantic Region.

Supplies and Services

In 1991, the cost of operating the Borden-Cape Tormentine ferry service was $37 million. In addition to the $24 million in wages and benefits, the regional economy also gained income through the purchase of $13 million in goods and services including, for example, food, fuel, paper products, cleaning agents, and repairs. More than a quarter of these purchases — $3.5 million — were bought directly from Prince Edward Island businesses. Purchases were made in all parts of the province. Approximately 70% of the total purchases of Marine Atlantic's Borden-Cape Tormentine service were from Island companies who had annual sales to the ferry company of less than $25,000. Three companies had contracts for over $100,000. Three others have sales to Marine Atlantic of between $50,000 and $100,000, and an additional 5 companies sell between $25,000 and $50,000 to the ferry company.

Informal interviews with a number of owners/managers of Island businesses resulted in a variety of responses concerning their loss of contracts with Marine Atlantic. Curiously, some large contract holders appeared to be unconcerned about the pending loss of income while smaller contract holders felt the loss would have a major effect on their business, especially in recessionary times. Small business owners also spoke of the uncertainty of the long-term benefits of a bridge for small business on Prince Edward Island.

Cafeteria and Gift Shops

Ferry terminal gift shops and cafeterias are included here because of their location within Marine Atlantic's ships and office terminals. All of them employ staff and generate income. Judson Foods, which operates a restaurant in each terminal building, employs 10-12 full time, and approximately 40 seasonal, employees. Purchases by Judson Foods amount to $250 million annually, and most of them are made in Prince Edward Island and New Brunswick. Bret Judson, owner of Judson Foods, said that these restaurants represent about 25% of his total business. The loss of this income, he said, would have a significant negative impact on his business.

The gift shops, both on the ferries and at each terminal, are owned by Earl J. Dimmock of New Brunswick. These shops provide 30 seasonal jobs, 35% of which are filled by students. The remaining 65% of employees use this seasonal work to qualify for unemployment insurance benefits, which provide year-round input to the local economies. Earl Dimmock said the loss of this income would have severe negative impacts on his business.

Economic Spin-off Effects

The economic activity created by wages and purchases in the local economy creates spin-off effects — other economic activities which capture the dollars now available in the area's economy, and create more employment. These are termed "indirect jobs." Annual inputs of close to $24 million to an economy the size of Prince Edward Island create large spin-off effects.

Using the accepted Statistics Canada model and multipliers developed by the PEI Department of Finance, it is estimated that the spin-off effect amounts to an additional 232 jobs created in the PEI economy. The $3.5 million that Marine Atlantic spends for supplies and services in Prince Edward Island generates another $2.2 million in income for the Island.

The economic impact of Marine Atlantic is of such significance that it should be thought of as a tool for economic development, not as a dispensable inconvenience. In a province with an unemployment rate in 1992 of 17.7%, 554 direct jobs plus the associated spin-off employment will not be easily replaced. A question raised at the Environmental Assessment Panel (EAP) hearings in 1990, concerning the loss of employment on the ferry service, resulted in this reply by Smith Green Management Consultants for Public Works Canada (PWC):

> The business community in the entire region expects that there will be long-term gains to the economy on both sides of the Strait. In view of the imprecision of the estimates of employment in all sectors, it is not possible to establish the extent to which new employment will offset the wage loss from the termination of ferry service employment.[2]

Marine Atlantic is one of the largest employers on Prince Edward Island outside the government sector. The largest private sector employer is Cavendish Farms, which employs approximately 600 workers. An unconfirmed estimate of their payroll is $13 million annually. By comparison, Marine Atlantic's Prince Edward Island ferry service employs 651 workers and pays out total annual wages of $19.2 million.

Clearly, the dismissal of Marine Atlantic's inputs to a depressed economy like that of PEI will result in major economic losses. To believe that these concrete losses will be replaced by "expected" but undefined "long-term gains to the economy" in the operations phase of the proposed project, is to cover one's eyes and hope beyond reason.

A Comparative Analysis

How will the economic void be filled in the post-Marine Atlantic phase of the proposed bridge construction? The precise lost dollar and lost job figures must be considered side by side with the imprecise economic benefits cited by proponents of bridge construction and operation.

Economic forecasting is fortune-telling-in-pinstripes, speculation masquerading as science. There is no better example of this than the duelling economic predictions associated with this specific bridge project. Two studies of both the construction and operations phases of a bridge across the Northumberland Strait are remarkable for upholding opposite forecasts of the impact of the project on the Prince Edward Island economy. "The Socio-Economic Impact of a Fixed Crossing to Prince Edward Island" was prepared for Public Works Canada in July, 1991 by the Atlantic Provinces Economic Council (APEC); and "Fixed Crossing Business Opportunities Study" was prepared for the Greater

Charlottetown and Summerside Areas Chambers of Commerce in January, 1992 by Coopers & Lybrand.

Specifically, the Coopers & Lybrand study for the Chambers of Commerce estimates that 35% of the total value of the project's goods and services could be supplied by Island residents and industry. The study identifies six areas of economic opportunity open to Prince Edward Islanders. However, prior to bidding on these contracts, upgrading or expansion of the Island's current capacity will be required in four of the six areas defined by the report: materials fabrication, facilities, equipment, and management and administration. In a fifth area, the supply of raw materials, the study says, PEI has little chance of winning many contracts. In one area alone is Prince Edward Island seen to be equipped to supply a significant component — construction labour. This is estimated to be, in total, $79 million of their base $750 million project total (10.5% of the total project).

Maximum economic benefits to PEI will occur in the bridge construction phase. And these benefits, says Coopers & Lybrand, are dependent on "one key factor — the location of the fabrication yards and head offices of the Project"[3] in Prince Edward Island. The report further recommends that "The Chambers should question the merits of the project proceeding if this condition cannot be met."[4]

Coopers & Lybrand recommend that Chamber of Commerce support for the bridge project should hinge on short-term business opportunities occuring during the construction phase, rather than any perceived benefit from, or even the necessity for, alterations to the Island's transportation infrastructure.

In contrast, a parallel economic study undertaken for Public Works Canada, was prepared by the Atlantic Provinces Economic Council (APEC) in July, 1991. Maurice Mandale, Senior Economist at the Atlantic Provinces Economic Council, is part of the team who authored "The Socio-Economic Impact of a Fixed Crossing to Prince Edward Island." In an interview, Maurice Mandale confirmed the pessimistic tone of the report with regard to construction phase benefits for PEI: "There won't be all that many." New Brunswick "stands to gain the most, as the natural location for staging and marshalling construction materials and supplies," the study said.[5]

Regional estimates of direct spending that could take place in the Maritimes range up to 77% of the total cost of the project. APEC's Senior Economist, however, is not confident that this economic activity will occur in the Maritimes, due in large part to private sector involvement in the project. Here he points to the experience at the Hibernia project in Newfoundland which had a much larger public spending component, "... there is substantial leakage from the region more or less at the whim of the principal private partners."[6] To maximize regional benefits,

Maurice Mandale says, "Public Works would have to wield a fairly big stick on it." And he is unsure if they would be able to do this. "Their [Public Works Canada's] hands may be reasonably tied in that respect because the successful bidder will have to have a certain regard for his own total costs." However, in contrast to the Coopers & Lybrand study, it is not in the winning of construction phase contracts that APEC envisions the benefits of the bridge. It is, rather, in the operations phase. And the benefits here are harder to nail down.

For the Atlantic Provinces Economic Council, the real advantage to construction of this piece of transportation infrastructure is in the effect it will have on enhanced productivity in the region.

> A final aspect of the APEC work is to assess the wider effects of building important pieces of infrastructure on the region's overall productivity. Work conducted in the United States over the past several years has indicated a strong link between physical infrastructure development and productivity. Where the former has been strong, as in Japan, the latter has increased faster.[7]

To Islanders, the necessity for better transportation infrastructure across the Northumberland Strait is undeniable. But for Maurice Mandale, a bridge is not a "necessary piece of work. It will, we think, add to the overall efficiency of the operations of the Prince Edward Island economy in that it removes a bottleneck in transportation." He adds that an improvement to the Island's transportation infrastructure across the Northumberland Strait does not, however, require construction of a bridge. Improvement in transportation infrastructure could also mean improved ferry service or a rail or road tunnel.

Once examined closely, the apparent economic benefits of a bridge melt away quickly. Bridge construction will terminate jobs at Marine Atlantic for 554 Islanders. It is questionable how much economic activity will occur on Prince Edward Island in the 5 years during which construction is to occur. And bridge construction is not necessary to improve the region's transportation infrastructure. Improved ferry service would serve the same purpose while maintaining or increasing long-term employment opportunities with a parallel enhancement to the Island's economy.

The economic downside — what Prince Edward Island will lose with the construction of a bridge and the loss of Marine Atlantic, one of the Island's largest employers — is the other side of the coin. Curiously, this has not been studied in any coherent way by the federal, provincial or municipal governments. This loss includes, on a yearly basis, $20 million in the high value-added spending of wages and benefits, $3.5 million in purchases made on Prince Edward Island, as well as 232 spin-off jobs. Further losses include the uncalculated increase in unemployment as a result of the projected out-migration of unemployed ferry workers and their families, and compensation/unemployment benefits. These losses have, to a major extent, been lost in the pro-bridge boosterism that has accompanied this project.[8]

Uncharted too, is the economic impact of the loss to PEI of versatility in its transportation sector. Currently, the Marine Atlantic ferry system is capable of carrying both road and rail traffic. Prince Edward Island lost its own rail service in 1989 after Erik Nielsen's Task Force on Program Review recommended abandonment. Considering the rebirth of rail transport in Europe and other parts of North America, it may have been a short sighted move on the part of the federal, Progressive Conservative government.[9] This seemingly myopic vision is compounded by the reality that building a road bridge binds Prince Edward Island irrefutably, for the 100 year life of the project, to road travel and truck transportation. Prince Edward Island will have sacrificed the versatility which it now possesses for a dangerously monopolistic and possibly outmoded system of transportation.

Transportation Strategy for Prince Edward Island

The trucking industry is expected to be the big winner if the proposed bridge is built. This industry has expressed consistent support for the bridge construction project. In a 1990 slide presentation to the EAP, Smith Green & Associates Inc. had this to say about the trucking industry:

> This industry, as the principle users of the ferries system, will experience the greatest impact from a fixed crossing. At an estimated saving of $63 per trip (154,000 trucks @ 1.8 hrs @ $35/hr), the industry would expect to save approximately $9.7 million per year if a bridge crossing were available.[10]

Proponents of the bridge frequently point to savings which may occur due to reduced transportation costs. But the question of just who would benefit from these savings is one which few reports are willing to tackle with any definitive answers. Smith Green says, "It is expected that some of these gains [savings] would be passed on to customers."[11] Maurice Mandale and the APEC study, too, talk of two varieties of transportation savings: the savings in time, and the savings in dollars which may occur with, for example, more efficient use of trucking equipment. A representative of truck operators addressing the federal government's Environmental Assessment Panel agrees with Maurice Mandale in saying that significant dollar savings will not be passed on to the consumer. The savings which will accrue to the consumer are savings in time.

Eldon Shorney of Midland Transport Ltd. addressed the EAP in March, 1990. He said, at that time, that Midland Transport Ltd., one of several trucking firms owned by the Irving family of New Brunswick, makes 16,000 crossings per year on the Borden-Cape Tormentine ferry system. "Time is money," Shorney said, and estimated that for the time wasted via ferry transport "the extra cost to Midland [is] ... over $600,000 per year."[12]

However, if savings in trucking costs accrue to the transportation companies, Prince Edward Island again will benefit only marginally. Of the trucking firms listed in the Inventory of General Freight Carriers, 1990 (published by the Atlantic Provinces Transportation Commission), 5% were registered to Prince Edward Island addresses and most of these companies employed fewer than 5 people. Only 3% of the total inventory were licensed to operate into New Brunswick — that is, only 3% would even have use for a bridge. If savings in trucking costs accrue to the transportation companies, Prince Edward Island's gain will again be marginal.

For the authors of the Gardner Pinfold report, just who would benefit if reduced transportation costs were to be realized, "would depend largely on competition in the trucking industry."[13] The Atlantic Provinces Transportation Commission, too, in response to a question from the EAP concerning the disposition of the estimated $10 million in annual anticipated savings, point to competition in the trucking industry to reduce rates and, in that way, pass on the dollar savings.

But, according to a 1990 report prepared by the Atlantic Provinces Transportation Commission (APTC), "Low rates, stiff competition and the dominance of the large trucking companies, has made it difficult for independent owner-operators to remain in business."[14] Indeed, stiff competition, low rates and the dominance of large trucking companies have been the subjects of an ongoing court case in New Brunswick concerning the concentration of Irving-family ownership in the trucking industry. Although the number of motor carriers licensed to operate in the Atlantic region grew by 70% between 1988 and 1991 (when deregulation in the trucking industry came into effect), the APTC reports that "the existence of a licence does not ensure that a given company has not gone bankrupt."

The APTC also states:

> ... deregulation has also allowed many new companies to compete in the market and force rates down to break even and, in some cases, below break even levels ... In the future, most carriers expect that, aside from niche players who specialize in particular products or services, smaller carriers will become less common as mergers and acquisitions create larger carriers. This will leave shippers with

fewer choices to carry their goods, and will likely result in higher rates.[15]

The consolidation in the trucking industry and the opinion of the motor carriers that it will "likely result in higher rates" and "few choices" should sound an alarm, particularly in Prince Edward Island. Although consolidation in the trucking industry affects New Brunswick and Nova Scotia too, they have retained the ability to ship by alternative modes of transportation, specifically rail. Though shipping by rail within Prince Edward Island is not currently possible, rail access to PEI remains intact — the rail line between Cape Tormentine and Moncton has been maintained for use during the construction of the proposed bridge. From there, rail access to Prince Edward Island can be maintained through the current ferry service or via a rail tunnel. As long as versatility in transportation across the Strait is maintained, Prince Edward Island will always have an alternative.

A road bridge to Prince Edward Island buckles the economy into a transportation straight jacket isolated for 100 years from alternative modes of that most vital of services, transportation.

Conclusion

During the ongoing debate about the fixed link, it has been difficult to maintain the focus of the discussion on investigating the mode of transportation that is best for the economic and social development of the Island. Most Islanders accept the fact that the choice of transportation link should represent a balance between convenience, economic efficiency, environmental limits and technology.

Although there have been hundreds of hours of debate and thousands of pages of documentation, there has been little attempt to consider all possible options for providing the most cost efficient, environmentally appropriate and effective transportation link. Only bridge proposals have been seriously considered. Proposals such as a road tunnel or a rail tunnel have been given only passing reference, even though they appear to be more environmentally acceptable and — the rail-tunnel at least — less costly. Improved ferry service has not even been discussed.

Interest in the project has revolved around two main themes: how to provide a mode of transportation between Prince Edward Island and the mainland which will promote the overall economic and social development of Prince Edward Island; and how to promote the construction of a fixed link as a way of jolting the economy of the Atlantic region out of its present doldrums. Analysts have speculated that the federal government's primary interest is in the short-term jobs that the construction of a bridge will bring, rather than in an effective transportation infrastructure or long-term development for the region.

It is true that a mega-project will stimulate parts of the Maritime economy in the short term. However, the long-term benefits to the Island economy are at best uncertain. The withdrawal of the ferries will cause significant losses. The construction and operation of a bridge will result in the transfer of the federal transportation subsidy from the public to the private sector for the next 35 years. As a result, tax dollars will provide profits to large private companies which will make little investment in Prince Edward Island beyond the initial construction period. If the transportation subsidy remains with Marine Atlantic, on the other hand, it will go into ferry workers' salaries and towards purchase of goods and services supplied to the ferries by Island companies. This represents a significant long-term investment in the PEI economy.

This study has demonstrated that Marine Atlantic makes a major impact on the economy of Prince Edward Island. But, apart from the economic impact, the use of a ferry service as a transportation link should be given serious consideration for other reasons. There is a perception that ferries represent an inconvenient and inefficient mode of transportation for both travellers and products. However, ferries provide a flexible and environmentally acceptable mode of transportation. The technology to make ferry services faster and more energy efficient is rapidly developing world-wide. The changes in ferry technology and other forms of transportation demonstrate that many countries are investigating systems of transportation that are not tied to road traffic. The Europeans, who are considered to be ahead of North Americans in transportation systems, are turning to more public transportation — which often means rail transportation.

Improvements to the ferry service must be given a high priority by the federal government. Following are some examples of proposed improvements: A key proposal of Marine Atlantic is the replacement of the *John Hamilton Gray* with a bigger ferry that would take a much larger number of tractor trailers. At present, the *Gray* takes only 14 tractor-trailers and, because the trucks have to back on to the ferry, more time is required to load and unload. In contrast, the *Abegweit* can accommodate 40 tractor-trailers. The proposed replacement ferry would take as many as 50 tractor-trailers. This will reduce the waiting time for both car and truck traffic. Changes could be made in the construction of the two summer ferries to allow them to transport passengers and cars more comfortably and economically until freeze-up. And there could be greater flexibility in accessing extra ferries in the summer to relieve the backlog.

In view of the results of this study, we strongly urge that a more detailed investigation be made of the possibility, the cost and the implications of improving the ferry service between Prince Edward Island and New Brunswick.

Transportation projects are too important to be manipulated only to provide short-term jobs. We must find the kind of transportation link that is flexible, cost-efficient, environmentally appropriate and effective for sustainable development on Prince Edward Island.

Notes

1. Marine Atlantic, *Annual Report*, 1991.

2. Smith Green & Associates Inc., "Response to Socio-Economic Issues Raised by Panel Experts." *Socio-Economic Slide Presentation to FEARO Assessment Panel*, April 1990, p. 11.

3. Coopers & Lybrand, "Fixed Crossing Business Opportunities Study." Prepared for the Greater Charlottetown Area Chambers of Commerce and Greater Summerside Chamber of Commerce, January 1992, p. 34.

4. Coopers & Lybrand, *op cit*, January 1992, p. 35.

5. Atlantic Provinces Economic Council, "Northumberland Strait Crossing Project: The Socio-Economic Impact of a Fixed Crossing to Prince Edward Island." Prepared for Public Works Canada, July 1991, p. 12.

6. Atlantic Provinces Economic Council, *op cit*, July 1991, p. 15.

7. Atlantic Provinces Economic Council, *op cit*, July 1991, p. 1.

8. See John Vanderkamp, "The Effect of Out-Migration on Regional Employment." *Canadian Journal of Economics*, no. 4, November 1970.

9. Atlantic Provinces Economic Council, *op cit*, July 1991, Appendix A, p. ii.

10. Smith Green & Associates Inc., "Socio-Economic Slide Presentation to FEARO Assessment Panel." Prepared for Public Works Canada, April 1990, p. 16.

11. Smith Green & Associates Inc., *op cit*, April 1990, p. 16.

12. Eldon Shorney, address to Environmental Assessment Panel. Summerside, March 23, 1990, 7 p.m. panel session.

13. Gardner Pinfold Consulting Economists Limited, "Northumberland Fixed Crossing Background Papers on Economic Impact During Construction and Operation." Prepared for Delcan/Stone & Webster, May 1989, p.2.

14. Atlantic Provinces Transportation Commission, "Impact of Transportation Legislation on the Atlantic Provinces for the Year 1990." Prepared for the National Transportation Agency of Canada, March 1991, Section II, p. 37.

15. Atlantic Provinces Transportation Commission, *op cit*, March 1990, Section II, p. 51.

Fears

"There doesn't seem to have been a strong popular demand for a fixed link crossing. The Islanders themselves have been quite ambivalent about this project. It seems to us that the project was initiated more as a business venture and not as a response to public demand. This suggests to us that business considerations will tend to mute the environmental considerations especially where the professional studies generally play down the risk. The lack of a strong popular demand for something like this also makes it harder to justify the inevitable environmental risks that we're incurring. And probably, in my judgement, explains the lack of sustained public focus on the project and its implications."

Mike Belliveau, Maritime Fisherman's Union,
at the Environmental Assessment Panel (EAP) hearings,
15 March 1990.

If anything has characterized the debate on environmental concerns associated with the fixed link it has been the duelling assessments among scientists of the effect the structure might have on the environment.

Part of the problem originates with a shortage of specific scientific studies, even basics such as accurate wind speeds over the Strait. And part of the problem comes from an absence of longer term baseline information from which environmental change can be quantified. Some people place their faith in scientific mastery and trust that science will solve all problems. Other scientists and environmentalists warn that the bridge, destined to be a "world's first" in several ways, is, if constructed, a huge and dangerous environmental experiment.

Ice specialists have variously predicted that ice will jam and block the western Strait, or will shear past the piers and leave the Strait within a few hours to a few days of normal ice-out times. Fishers, scientists and environmentalists fear that the bridge's 68 massive piers will hold the ice for weeks longer each spring, lowering water and air temperatures, affecting farming and the fishery and the livelihood of thousands of farmers, fishers and plant workers who live on both sides of the Northumberland Strait. Other scientists assured us that risks could be "mitigated," and this became a buzz word to gloss over bothersome details.

The problem is epitomized by the concerns of a retired engineer in Charlottetown who has also worked on the ferry boats. Maurice Lodge's concern with ice goes beyond jamming at the bridge piers — he thinks there is a risk that the structure cannot withstand what he's seen the ice in the Northumberland Strait do. His view isn't popular and is only spoken of by others in hushed tones, for fear of betraying a lack of confidence in modern technology and science's ability to "mitigate," to overcome all problems. But he's not alone in his concern.

The possible environmental consequences of this megaproject are both chilling and unnecessary. After construction is completed and corporate Canada has tucked in its last vested interest and gone home, it is Islanders and Maritimers on both sides of the Strait who will be paying for environmental degradation, long after the bridge itself becomes obsolete — or comes thundering down.

Environmental Consequences of a Bridge

Irené Novaczek

"Development can never benefit more than a minority; it demands the destruction of the environment and of peoples."
The Ecologist, July / August 1992

There is no doubt about it — the proposed bridge between PEI and New Brunswick is a mega-project, and mega-projects generally trigger unforseen and unpredictable environmental change. In the case of this fixed link, there are more outstanding questions than there are answers and the uncertainty regarding possible consequences has produced tremendous controversy.

One thing we *can* count on is that the bridge will be built at tremendous cost, even if you consider only the environmental degradation and energy consumption required to amass, transport, transform and put in place the quarried rock, the smelted steel and the cement. Rock quarries all over the western world are subject to protest, as the social, medical and environmental costs of this activity are increasingly recognized. Because workers and local residents exposed to the dust that comes with quarrying often suffer serious lung disease and painful and untimely death, and also because of the attendant noise and destruction of landscape, this industry is no longer welcome in many communities. In Atlantic Canada, many communities are battling even small quarries. The potential impact of establishing one of the world's largest quarries on the Miq'mak peoples' sacred mountain, known commonly as Kelly's Mountain in Cape Breton, is staggering. If this quarry goes ahead, it may well be in part because of the prospect of business from the PEI bridge construction, and it will destroy not only lives, fisheries and tourism opportunities, but also the largest tract of rare, unspoiled wilderness on Cape Breton Island.

Iron and coal mining and smelting of steel are also now recognized as threats to our precious remnants of wilderness and to our surface and ground water resources. The massive concrete fabrication yards required for mega-project construction also exact their toll. Concrete fabrication involves the introduction into the air of lung-destroying dust, the consumption of massive amounts of groundwater, and the release of environmentally dangerous wastewater.

The time has come for the careful assessment of any project that involves the large scale environmental destruction such as that which comes with building immense concrete and steel structures. The proposed bridge between New Brunswick and Prince Edward Island is not simply an opportunity to put people to work. It must be seen in the clear and rational light that exposes all of the long-term social, health and environmental costs.

Problems with the Bridge

It cannot be denied that conditions in the Northumberland Strait will change as a consequence of this project. Localized damage resulting from the dredging of the bottom and the placement of bridge pilings, the risk of accidents involving the increased levels of marine traffic, the disturbance to wildlife from the noise and pollution that accompany construction are just the bare beginnings. Much more disturbing are the long-term ramifications once the bridge is in place. Where there once was open water there will now be a bridge, and this must have an effect on the movements of water, longshore sediment, ice and, possibly, on migratory fishes. The structure will pose a hazard to surface shipping traffic and provide increased opportunities for spills of hazardous chemicals into the marine environment, both from road accidents and shipping accidents. The bridge's superstructure will be a hazard for small birds in flight, mesmerized by the night lights. Conditions will deteriorate in Cape Jourimain Wildlife Sanctuary and Noonan's Marsh, which will be traversed by the approach roads and spans. Nesting, feeding and migrating birds will be disturbed and, over time, particulate fallout from the exhaust of passing traffic will poison the marshes and their wildlife.

The question is, how great will these changes be? Some would argue that this construction will be the saving of the economy and a world-class tourist attraction. Others contend that it will be another blight on the face of the earth, and an ill-conceived plan that will result in economic, social and environmental ruin. Will the fishery of the Northumberland Strait be consigned to history? Or will the bridge be benign and simply provide increased rocky bottom habitat for the lobster fishery? Will the bridge piers slice the winter ice floes into slivers? Or will it hold them back, choking the Strait and piling up behind the nearshore piers, grinding the bottom to dust?

Whether or not the bridge will interfere with the movement of ice out of the Strait in spring is one of the major unresolved controversies. Bridge proponents are proceeding on the assumption that there will not be any significant problem. They base this assurance on a computer modelling exercise, but do not admit to the limitations of their model. The model has been criticized both by scientists from the Department of Fisheries and Oceans (DFO) and other experts because it is based on very limited data, much of it extrapolated rather than "real," because it does not take into account the peculiarities of the tidal regime in the Strait, and because it has never been tested for predictive accuracy against the reality of ice conditions in the Strait. This is the shaky ground upon which critical decisions have been made. The long-term impact could be felt throughout the southern Gulf of Saint Lawrence in terms of reduced productivity and depressed fisheries.

Our experience with the Canso Causeway is a warning that we should prepare for the unforseen. When that structure was proposed, even DFO scientists were confident that it would have little negative impact. Immediately following construction, it seemed that they were right, but now we know better. Over the years, major losses in the lobster and herring fisheries near the causeway, and possibly losses in the Cape Breton gaspereau fishery, have been attributed to this interference with water movement in what appeared to be a minor waterway. Agriculture has also suffered through the significant reduction in the frost-free season. None of these effects were predicted.

Although the currently proposed bridge has been designed to mitigate some of the predictable negative effects, many, including ice scour and potential interference with tides and currents, can never be designed away. Every bridge has piers which effectively narrow the channel, in this case by about 6% at the waterline and 12% at the bottom. Also, no design can avoid relatively close placement of piers in shallow, near-shore water. With piers so close together, build-up of ice is inevitable at least in the shallows, and this will result in increased scouring and destruction of valuable bottom habitat. The area west of the bridge is a herring spawning area. The herring spawn only on marine plants, and these plants are confined to small patches in shallow water. This is exactly the habitat most at risk from increased ice scour. Potential changes to tidal and other current patterns, and the effect of these on biological systems, are very difficult to predict because we know so little about dynamics in the Strait. However, DFO scientists have pointed out the potential for higher tides, stronger currents, increased coastal erosion and sedimentation of sensitive habitats. Changes in tidal and residual currents, or to gyre systems that may operate in the Strait, will have complex and unpredictable effects upon the life history and biology of many life forms, including commercially important species. No one knows what the long-term results will be, nor can anyone say how far away from the bridge the effects may be felt.

Bridge proponents contend that this project has been studied more extensively than any other project in the history of Canada. Given the brief history of environmental assessment in Canada, this may be so.

However, quantity of paper output and quality of science are two different things. Respected scientists from DFO and universities have gone on record as having doubts as to the adequacy of data, and have also pointed out questionable methodologies and hypotheses evident in the consultant's reports produced on behalf of Public Works Canada (PWC). Research work promised by PWC, that might have filled the gaping holes in the knowledge base, has never been performed. As a member of the review committee trying to define studies that would allow Strait Crossing Inc. (SCI) to monitor environmental effects of construction and operation, I am seriously concerned about the lack of basic knowledge of how different organisms interact with each other and with the ice, water temperature, tides and currents. This lack of information severely restricts the potential for in-depth effects monitoring. There is a real danger that any monitoring exercise will fail to detect important changes that will have long-range impacts upon the Northumberland Strait.

Unfortunately, there is no bridge in the world that we can study to shed light on potential long-term effects. The Northumberland Strait is unique in terms of tidal currents and seasonal ice cover. Even if another bridge did exist in a similar environment, there would be no baseline data available, and no history of environmental effects monitoring to study. Previously built bridges have simply not been thoroughly scrutinized from an environmental perspective.

Impact on the Ecology of the Island

Many grave concerns were raised by the Environmental Assessment Panel (EAP) that studied, and rejected, the proposal for this bridge. Among them was the problem of the unknown potential carrying capacity of the Island. An island, by its nature, has limited resources. Our island, in particular, is fragile because of its easily erodible soils and fractured sandstone bedrock. We have very little surface water, and easily-contaminated groundwater. There is limited coastal land suitable for development, limited ability to absorb air pollution, limited ability to deal with increased traffic, increased garbage etc.

It is commonly perceived that more tourism, for instance, would be beneficial to the economy. This may be true, but we have no idea how much will be too much, or whether the type and number of tourists who will be attracted by a bridge, as opposed to a leisurely ferry ride, will be a blessing or a curse. At what point will the best-loved beaches and attractions become degraded, over-crowded places to avoid at all costs? (Some would say this is already the case.) At what point will the roads become clogged, and roadsides hopelessly polluted? (Again, there are already concerns being expressed by residents.) At what point will the estuaries be so ripe with sewage that aquaculture and fisheries will be ruined? (Many estuaries are already closed for shellfish harvesting.) At what point will we no longer have room for the fast food containers, beer cans and disposable diapers that come with tourism? (Our landfills are already a problem.) Until we have stopped to consider our limitations, it is wise to proceed with great caution. The Island already faces serious challenges in the form of groundwater contamination in the most built-up areas, soil erosion and pollution of inshore marine waters. Without careful study and planning, a headlong rush for increased development will spell even more rapid deterioration for the Island's ecology. Beyond a point, the Island will not only be uncomfortable for residents, but will lose its attractiveness for tourists also. A sustainable future for resident Islanders will not come in the form of 10-week summer tourist jobs. It is and always will be dependent upon healthy soil, pure air and water and clean rivers and bays, to support our primary industries: agriculture and fisheries.

The Link and Global Trends in Transportation

At a time when Canada has made an international commitment to the reduction of greenhouse gas emissions and more progressive countries are looking seriously at upgrading train networks because they are so much more fuel efficient for long-range transportation, Canada is building an $840 million 2-lane road bridge to last more than 100 years. Out of necessity, we must develop more rational and less damaging means of transportation within that 100 years, and a bridge may not be adaptable to future needs. As well, in the next 100 years, rising sea levels and increased coastal erosion could dissect PEI into several small islands, leaving us with a bridge to virtually nowhere, and a pressing need for ferries.

Federal Environmental Assessment

Unlike those good old/bad old days, when development reigned supreme and environmental effects were not even considered, much less monitored, Canada now has a web of environmental laws and a legal review process for judging the potential risks of any proposed development that involves federal money. The process under which original data is gathered on the environment that will be affected by a project is called an Environmental Impact Assessment (EIA). For projects deemed to have major or unknown negative effects, or around which there is significant public concern, the Federal Environmental Assessment and Review Office (FEARO) puts together a panel of independent people who have relevant expertise, to consider the project in detail — this panel is called an Environmental Assessment Panel (EAP). The panel considers the EIA, listens to expert witnesses, holds public forums, asks for further studies as required, and then makes a recommendation. This process is called an Environmental Assessment and Review Process (or EARP).

The law requiring environmental assessments was established in 1973. Unfortunately, this law is in the form of a guideline rather than a regulation. Projects go through an in-house FEARO screening procedure, at which time the decision is made to proceed immediately, perform some degree of impact study (EIA) or go to full public review (EARP). Of every 1,000 projects, only 10% undergo further study and 1 of the 1,000 goes to an Environmental Assessment Panel (Weston 1991).

In 1992 the law governing environmental assessment was revised and this should come into force in 1993. Under the amended law, an EAP must carefully consider not just the potential impact of the proposed development, but also the impact of all reasonable alternatives. In other words, the Panel has to be convinced that the project is not only necessary, but is also the best means of accomplishing the desired end. An EAP convened now to consider the PEI-N.B. bridge would have to consider both a rail tunnel and an improved ferry service, and would doubtless conclude that a bridge was the most risky alternative. However, even the new law still has a fatal flaw. The ultimate decision on whether a proposed development project is unacceptably dangerous to the environment is not given to the panel of experts brought together to weigh the evidence. The final judgement is political, and the federal minister in charge can choose to reject or ignore the findings of an EAP.

In the case of the EAP convened to consider the fixed link proposal, an already flawed process was made even less effective by the timing chosen by the federal government. The Panel was not convened at the very early stage when it could have considered the relative risks and benefits of the various possible solutions to the transportation problem of PEI. The question was not: What is the best option for improving transportation services? The option of a tunnel was rejected before the Panel was struck, basically because only proposals for a road tunnel (not a more benign rail tunnel) were called for. The option of improving the ferry service was never contemplated. The Panel was asked to study only the option of bridge construction. However, the timing was such that they had to consider the *concept* of a bridge rather than any particular design. This left the process open to subversion and abuse.

After deliberating for over a year, poring over detailed documentation and scientific studies, and listening to public opinion as well as expert advice, the EAP concluded that there were too many known and unknown risks in the bridge proposal, and advised that the project should not proceed. In particular, after hearing from many concerned and qualified scientists from within DFO as well as from academic circles, the Panel concluded that we simply do not know enough about the Northumberland Strait to meddle with the system. Further, it would take years of careful study to appreciate the range and causes of natural variations in such things as growth and reproduction of commercially important species and their food sources. The Panel urged that, prior to any further action, appropriate studies and data gathering should be carried out. Although years have elapsed since these recommendations, and in spite of the apparent determination on the part of the federal government to proceed, no such studies have been initiated. A panel of ice experts was convened to rework the same data base in light of a specific bridge design. For this exercise, no new data were collected. The controversial computer model the ice panel worked with was never field-tested to see if it could accurately predict the timing of ice-out. Apart from the ice modelling, no attempt was made to address the many other problems identified by the EAP.

Because the EAP did not have the chance to evaluate a particular bridge design, the way was open for pro-bridge parties to claim that, although the generic bridge was unacceptable, the latest design would be much safer and therefore acceptable. It has been repeatedly claimed that the current design addresses all of the environmental concerns raised by the EAP. But ... the original Panel has never been reconvened to consider the facts and determine whether this is indeed the case.

The Argument for Another Assessment

Shortly after Premier Ghiz signed the agreement that cleared the way for further negotiations between the bridge construction company, Strait Crossing Inc. (SCI), and the federal government, the anti-bridge coalition, Friends of the Island, launched a court case to challenge the environmental review process for the proposed bridge. The argument was that, by convening the EAP to consider a generic concept, and then proceeding with a particular design, the government ignored the spirit of their own law, violated constitutional principle, and set a dangerous precedent for the future of Canada. Indeed, if this type of process is to be allowed, there is a good chance that no development project, no matter how inappropriate, could ever be stopped on environmental grounds. The government could continue to convene any panel to review a concept, and then, if the concept did not gain acceptance, simply pull out a particular design that appeared to "take care of" the panel's concerns, and proceed without any further public study. The court case is, therefore, viewed by environmentalists as a landmark case, which tests whether the federal government is committed to environmental protection, or just interested in the appearance of due process. As of this writing, Friends of the Island successfully challenged Public Works Canada in court on these issues. PWC is proposing to appeal the court's decision, which ordered further environmental assessment.

Public Involvement in Environmental Decision-making

The public in 1993 is sophisticated, knowledgeable and concerned to a degree that is new for this country. People need to know, deserve to know and demand to know what sacrifices they will be making in exchange for economic development. The public is no longer mollified by a political or business leader saying "trust me."

However, when economic times are hard and people are trapped in mortgages, car payments and debts accumulated in the pursuit of what goes for the "good life" in modern society, irresponsible developers can stir up a frenzy of support for projects offering short-term economic relief. Under this type of pressure, considered critique of environmental concerns goes out the window. It does not help when there is no detailed information presented in public upon which people can make a rational judgement. That is what happened in January of 1993, when SCI held a series of public meetings that were ostensibly to provide opportunities for the public to critique the draft of an environmental management plan for the bridge.

The written document circulated in advance of the meetings held considerable detail on land-based construction procedures. Unfortunately, for those looking for details on plans for monitoring and studying wildlife and marine habitats, there was nothing but vague generalities. Lists of species that could be considered for study were given, but there was no indication of which were considered essential or important. No planned procedures were outlined, no personnel or equipment or techniques were presented, no schedule of actual studies was given. Even the geographic limits of the study area and any mention of a potential budget were lacking. The public meetings offered no more information. What the public actually heard at these meetings (I attended 3 of the 4 on the Island) had little to do with environmental protection and a great deal to do with jobs, jobs, jobs. The presentation varied with the locality, in that it highlighted economic benefits to the particular town hosting the meeting. Comment elicited was, not surprisingly, dominated by questions about jobs, followed at a distance by questions about compensation packages and, running a barely perceptible third, questions regarding the environmental management plan. Legitimate critique of environmental impacts of the bridge was handily side-stepped, and the public was left no wiser concerning the details of critical items such as the planned baseline studies and environmental effects monitoring.

The risk lies in the fact that we are still largely ignorant of what is in the Strait and how the various elements work together as a system. We do know that physical conditions, such as the timing of ice-out, vary greatly from year to year. To gain any appreciation for the complex patterns of change through the seasons and from year to year would take many years of study, and we no longer have the luxury of time. Current plans are to proceed with construction after less than a year of baseline data collection, and then to carry on with development of protection strategies and with the monitoring of a few of the many, possibly unpredictable, environmental effects. It will be a grand attempt to perform credible science on a construction schedule, and even with the best of intentions, it cannot be adequate. Because the bridge is to be built by a private company, with deadlines to meet that will cost money to ignore, it is unlikely that any phase of construction will wait for the completion of detailed scientific study.

Once the SCI bridge proposal is accepted by government, the construction firm cannot be voted out of office. They will be acting under a legal contract and, barring immediate catastrophic results, the bridge will be built as scheduled, regardless of any change in public opinion or political control.

In a world where the end of resources of fossil fuel, virgin forests, clean air and fresh water are now in sight, where thousands of species of living things disappear every year, and where the human population is rapidly rising, further development in the traditional sense is suicidal folly. The proposed bridge between Prince Edward Island and New Brunswick represents a dangerously old-fashioned approach to economic development: the pursuit of possible short-term economic gain by means that threaten not only the physical and biological environment, but also the people in coastal communities who depend on the Northumberland Strait for their livelihood.

References

Weston, S.M., "The Canadian Federal Environmental Assessment and Review Process: An Analysis of the Initial Assessment Phase," 1991, Minister of Supply and Services Canada, Cat. No. EN 1007-3/21-1992, 67 pp.

The Effects of Climate and Weather on a Bridge

Owen Hertzman

If you take the Cape Tormentine-Borden ferry on a sunny summer afternoon, thoughts of the effects of climate on a bridge in the same area are probably very far from your mind. However, if you've been waiting for a ferry for hours on a summer weekend, with small children or a truckload of goods, you may have wondered what was so terrible about the idea of a bridge.

I am writing about the effect of the proposed bridge on climate in nearby areas, about weather effects on the bridge, and about the process used to evaluate these effects. To understand these issues we have to understand a few things about climate in general, and about what controls it in the PEI area.

Climate and Weather in Atlantic Canada

Climate is defined as the weather in a certain place, averaged over a certain time period. The weather is defined as the temperature, precipitation, wind and sky conditions observed at a particular place at a particular time. Weather is observed hourly, and more frequently during storms. Climate is usually described using daily information which is averaged weekly or monthly. Ordinarily, most people assume that the climate in their area varies only from month to month through the seasons — that variation is clearly the largest variability. For the bulk of this discussion, we will make the same assumption — that climate expressed by 30 year averages is "constant" except for seasonal variations.

When we examine the climate of a region we must examine both what happens and why it happens. The "what" includes not just the average "climatic conditions" for, say, the last week of January, but some measure of the variability of the weather from year to year, which is averaged together to determine the climate. The "why" of climate includes descriptions of weather patterns and other conditions which cause changes from season to season, and between years.

In Atlantic Canada the climate is dominated by two very different regions upstream: central Canada and central coastal United States. Air originating in these two regions is quite different, particularly in winter. Winter continental air from central Canada tends to be cold and dry, while winter maritime air from Cape Hatteras, New Jersey and Cape Cod tends to be moist and cool. Winter precipitation tends to be associated with days that are warmer than average. The average wind direction in the cold season is from the northwest. The strong flow from this direction is associated with the cold outbreaks which often occur after the passage of winter storms.

In Atlantic Canada in the summer, warm, humid air from the southeast U.S. occasionally reaches the region. Most of this air is modified by and mixed with cooler, drier air from regions further north before reaching southern Nova Scotia or western New Brunswick. Thus, PEI rarely experiences this air in unmodified form. Much of the air in the summer comes from the southwest since the average wind in the warm season is from that direction. Summer precipitation values are less, on average, than winter values, with rain falling most often on days that are colder than average.

Local Climatic Effects

The important control in the climate system in PEI is the state of the ocean which surrounds the island — whether it is in the form of ice or water; the temperature of the water; and, if the ocean is in the form of ice, how continuous the ice is. Elsewhere in this volume you will read in detail about the effects of the proposed bridge on ice in the Northumberland Strait. Here, I will make some more general comments about the interaction between the sea and the climate.

The salient fact about the climate in Atlantic Canada is that there is a thermal lag behind the solar determined seasons. That's a fancy way of saying that the onset of spring and summer are delayed by something cold and the onset of fall and winter are delayed by something warm. That "something" is the Sea Surface Temperature (SST). If one compares the climate of Charlottetown (the PEI climate station with the longest, best data) with that of the west coast of Canada, the biggest differences occur in the winter months. Thus, the climate penalty (shorter growing season, higher heating bills etc.) is really confined to the cold part of the year. If one compares the transition months in spring and fall (such as March/April versus September/October) the difference in monthly mean temperature at Charlottetown is more than 11.2°C, with the fall being warmer. At Vancouver, a location with no sea ice and deeper water, the difference is about 4.8°C. So we say that the thermal lag is larger in PEI. Why should this be so? Basically, shallower water and colder air brought in during the winter favour the initiation and growth of sea ice in the Northumberland Strait and the Gulf of St. Lawrence, but not in the Georgia Strait near Vancouver.

In the winter, ice typically begins to form near PEI in December or January. In the spring, the Gulf of St. Lawrence and the Northumberland Strait usually contain ice until May. The fact that both of these water bodies are very shallow enhances their ability to form ice in the winter and to warm rather markedly in summer (up to about 16°C in the Gulf on the north side of PEI, and up to 18-20°C in the Strait on the south coast).

The effect of the sea ice is threefold: It enhances cooling locally in coastal areas; it enhances the chance of fog; and it increases the chance that precipitation will fall as snow rather than rain. Each of the latter two conditions contribute to additional cooling in the spring because of disruption of the main heating source — the incoming solar radiation. Fog acts like a thermal insulator for the region when the average days are shorter than the nights and the infrared cooling exceeds the solar heating (as in the fall). But in the spring the high reflectivity of the top of the fog keeps the surface below colder than it would have been without the fog. The chance of fog even in winter is enhanced because of the open water in the Bay of Fundy and among the sea-ice floes.

The snow acts in a similar way to fog with respect to the sun, increasing the reflectivity and decreasing the amount of solar radiation absorbed by the ground (or the remaining sea ice) for a few hours to a few days, depending on the amount of snow and the subsequent weather conditions. Thus, the ground warms more slowly and "remembers" the snowfall thermally, sometimes for weeks. In addition, a heavy snowfall can actually add to the amount of sea ice, through thawing and re-freezing. Any change in the amount of ice can have an effect on the climate over a region which is at least as large as the area that contains the changed ice amounts. Thus, rather small changes in the fraction of ice in the Gulf or the Strait could change the weather near Summerside or Charlottetown quite significantly. One of the shortcomings of the environmental work done for the bridge proposal is that detailed examinations of these scenarios have not, so far, been made.

Now, how does the ice directly change the weather during the spring in the bridge area? During many storms, the low-level air flow (the wind direction near the surface) is from the northeast. Thus, increased amounts of ice in the Strait will lead, via sea-air exchange, to colder air coming over eastern New Brunswick and northern Nova Scotia. This colder air can change the composition of rain so that it contains significant amounts of freezing rain, ice pellets or snow. It also can increase direct cooling of the ground. There is a similar relationship between the Gulf of St. Lawrence and PEI. Between storms during the spring, the mean wind direction is from the southwest, so the conditions in the Strait and on the mainland affect PEI. In short, if an area nearby is cooled, eventually that cooling will affect your location.

So, to summarize, the climate system in PEI has both large scale influences typical of Atlantic Canada in general, and local influences because of the characteristics of the surrounding ocean. Now let's look at the effect of the bridge in all this.

The Bridge and Current Climate

The direct effect of the bridge structure on the climate should be negligible except in the area very close to the structure itself (perhaps a few hundred metres to a few kilometres). The structure itself is small enough in area that direct effects on the solar radiation should be small. Wind changes, and resulting precipitation changes, should be very local. However, effects on the climate because of changes in sea ice might be quite substantial and extend perhaps 100-200 kilometres from the bridge in all directions.

The changes would occur as follows. At the start of the ice year, the presence of the bridge, and the absence of the ferries, would allow sea ice to form earlier, and grow more quickly. The average afternoon temperature in December is close to 0°C while the average daily temperature is about -4°C. If ice begins to form earlier because of the bridge piers and the lack of the icebreaking action of the ferries, a portion of the western part of the Northumberland Strait would cool more rapidly with the bridge than without. Any marginal rain-snow event occurring during this "freezing" season would have a higher probability of being snow. This would enhance the growth of more ice, decrease solar energy into nearby land areas and increase the chance that the next storm will also have a higher fraction of snow, and have colder winds near the surface. Thus, it seems clear that the start of the ice season will be earlier if the bridge is built.

During the ice season the bridge will increase the chance of ice jams in the Northumberland Strait. There has been much argument over how much of an effect this would have, with the proponents and opponents of the bridge arguing over what fraction of the channel could be blocked, how large the ice floes would be, and how much icebreaking the bridge itself will do. However, consider that the ice climatology of the whole region is likely to be changed by the bridge because of the effects at the start of the ice year. All other external factors being equal, there should be more ice on, say, February 1 with a bridge than without one. Suppose we have 8/10 regional ice cover at this time instead of 7/10. We then have enough open water to provide additional moisture to a storm coming up the Atlantic coast. We know that most of the storm's moisture comes with it, but the local moisture can enhance precipitation rates and increase snow showers after the main storm has passed. If we have 10% enhancement of total precipitation and 10% higher snow fraction because of the presence of bridge, for a 40 mm water equivalent total event we would expect an additional 8 cm of snow on land — equivalent to perhaps 1-2 cm of extra ice in the Strait and Gulf. For this to occur at any one place is not a problem. However, if this occurs over the entire Strait area during the entire season, there is the potential for significant changes in ice amounts and thicknesses during the main ice season.

During periodic warm periods in winter, what will happen? The land area surrounding a bridge will have more snow than it would have received if there was no bridge. Therefore, warm low-level southwesterly winds will be cooled and moistened, increasing the chance of local snow showers and decreasing the heating from these winds. Thus, thawing of snow and sea ice will be retarded.

What about the end of the ice season, the melting period in late March, April and May? Obviously, if we start with more ice it will take longer to melt, since there is more substance to be melted. However, the rain-snow and snow-reflectivity mechanisms are still in effect, though with decreased efficiency because of the warmer days. Extra snow in April or May slows down the warming cycle, perhaps for a few hours if it's only 1 or 2 cm, or perhaps for a few days if it's 10-15 cm. The key point is that the effect is similar to the effects that previously occured earlier in the year.

So, if the bridge increases the amount of sea ice in the Strait, and the amount of sea ice increases the chance of snow versus rain and colder low level winds, then the amount of sea ice should increase at every stage of the ice season. It is unlikely that these effects, which have not been considered quantitatively and in an organized way by the proponents of the bridge, are within the guidelines set out by the Environmental Assessment Panel (EAP) — they have found that a delay in ice-out of more than 2 days would be acceptable if it occurred only once in 100 years.

The Possible Interaction with Climate Change

All of the above discussion assumes that the climate, expressed as averages, will not change over the next several decades, the presumed lifetime of the bridge. But what if there are changes over these "secular" time intervals? If the net effect is warming by 2-3°C, both globally and regionally and in all seasons, then any negative effects of the bridge (ie. delayed ice-out) are likely to be overshadowed. Under this scenario it is possible that significant ice might not be a factor in the area at all.

However, if the effect of global warming is weaker than predicted and variable from year to year, it is possible that the amount of sea ice in the Gulf/Strait area might increase rather than decrease in some future years. Under essentially all Global Warming scenarios there will be increased runoff from the continents to the Arctic Ocean and increased "fresh water" flowing from the Arctic Ocean into the Atlantic Ocean. (By "fresh water" I mean salt water with a lower salinity equivalent to an amount of fresh water mixed with "expected" sea water). This "fresh water" can enter our discussion in two ways. First, it will cool off the regions through which it passes. Second, it will slightly increase the probability of freezing because of decreased salinity.

Some of this Arctic water will enter the cold Labrador Current and cool the ocean near Labrador, Newfoundland and northern Nova Scotia. Some cold water will enter the Gulf of St. Lawrence through the Strait of Belle Isle. The effects on the regional climate are a regional cooling via sea-air exchange and a possible increase in the length of the ice season. Working in the opposite direction are the warming effects of warmer air over central North America and warmer water in the St. Lawrence.

In some years the regional cooling effects are likely to be weaker than the global warming effects. In these years, the bridge's effect on the sea ice is rather unimportant since the ice-out date should be very early — early enough that the control on marine biological activity is more related to solar radiation levels than amounts of sea ice. However, in the years when the regional cooling effects are important, the same mechanisms discussed above for the no climate change scenario come into play, with the following changes: If the bridge helps to initiate ice formation earlier than expected, the actual amounts of precipitation in the region would likely be larger, given the general warming of the regions upstream, such as the southeast U.S. This increased precipitation amount would enhance the effect of the bridge on the sea ice. In fact, any structure or other factor which increased ice locally could have a larger effect relative to the no bridge case, *even if the actual amounts of ice are lower than the current climate scenario*. Since the EAP guidelines do not specifically address climate change, presumably the comparison of any future changed (global) climate scenario with a bridge must be compared to the same (global) climate without a bridge. In that case, there are likely to be a few years out of the next 100 where this situation applies and where the likely negative effect of the bridge could be profound.

Weather Effects on the Bridge

To drive safely anywhere, one requires three things: a good driving surface, good visibility, and wind conditions which do not move one's vehicle significantly. In this section I will illustrate that the proposed bridge has not been demonstrated to provide these three safe driving elements.

At the time of writing, the public has not seen actual designs of what the Northumberland Strait bridge will look like to a driver in a moving vehicle on the structure. The actual design of the bridge is important in determining the effects of wind, visibility, and possible surface icing. It appears that any bridge structure which provides a safe driving environment in periods of high winds and/or intense snow would afford very little view during the crossing. Such a bridge would actually resemble an elevated tunnel. Since the structure is being sold to us partly as a tourist enhancing facility, this fact acts against the bridge in any cost-benefit accounting. In addition, keeping the proposed structure clear of ice may compromise the integrity of the surface and/or require enhanced maintenance personnel, equipment, and other costs which have not been anticipated in the current proposal's financial plan.

Wind Loading and Safety

Driving in high winds on high bridges with inadequate barriers is not pleasant. In western North America, a region with frequent high winds, but no sea ice and relatively little icing due to freezing rain, short bridges in metropolitan Vancouver, Seattle and Portland offer driving challenges, even relatively close to shore. In at least three places, floating bridges have been built instead of high bridges in ice-free zones, in order to avoid some wind-related problems.

The bridge proposed by Strait Crossing Inc. (SCI) has a driving surface 40-64m above water level. During all but very calm conditions there is a great deal more variability in wind speed at these elevations than at ground level. This is due to the presence of eddies, or rotors of air, which bring fast moving air from the upper parts of the atmospheric boundary layer closer to the surface, and slower-moving air up from near the ground. The air 1-2 kilometres up from the ground can be coming from a quite different direction than the surface air, and have a substantially higher speed — perhaps double, in some cases.

When comparing existing measurements done at 3 or 10m above the surface with the conditions on the bridge, assumptions are made about the average conditions at different heights under high wind conditions (neutral stability wind profiles). For most applications, these estimates are quite useful — they are certainly useful in comparing measurements made at different heights above the surface to discover spatial wind patterns for climate studies. However, when one asks about driving conditions on a bridge over periods of 10-15 minutes, other factors must be included. The "gustiness," or the variability, of the wind speed and direction must be taken into account. (For example, if wind conditions are coded as 0435G60 by an observer this means winds are from the northeast at 65 kph, with peak gusts over a 1-2 minute period of up to 111 kph.) Now, in designing a 1-2 kilometre bridge, one could use the 65 kph average wind and assume that a 111 kph gust would be rare for an individual driver. For a long bridge, the 111 kph gust is quite likely to occur, perhaps more than once for the same driver. So the wind effects on the real driving conditions on the longer bridge are much more serious. If slow convoys are required because of simultaneous visibility and/or icing conditions, the number of high wind gusts per vehicle increases.

Public Works Canada (PWC) commissioned studies which claim to show the number of hours per month when there will be wind related delays on the structure. These delays are attributed to reduced speeds and/or convoys required by the high wind conditions. In addition to not factoring in the "gustiness" problem outlined above, those studies have been done for the "generic" bridge design, not the actual SCI proposal. The studies have limited relevance to the particular proposal being considered now. Until such studies are done, the bridge must be considered of unproven safety in the wind conditions of this region.

Visibility and Safety

Among those familiar with weather conditions at sea in this region in winter, including local oceanographers and meteorologists, reactions to driving across the proposed bridge were rather negative. Some said they would choose not to drive it in winter. They pointed to the gustiness of the winds in the region and the sudden snow squalls which often characterize onshore flow (from water towards the land) during this season.

When one talks of the SCI crossing, one is discussing a 2-lane structure, with no turn outs. In the winter there are prolonged periods of high winds before, during and, especially, after winter storms. The wind speeds during these events often reach sustained 65 kph with gusts to over 92 kph. The post-storm surface winds are typically 28-37 kph from the northwest, approximately perpendicular to the direction of the bridge span, with frequent higher gusts during convective outbreaks.

The winds before a storm are often associated with rising temperatures, thawing and regions of light precipitation interspersed with times of heavier showers. Wind speeds can often rise quite rapidly from near-calm values. During the storm, winds can be intense, gusty and highly variable in direction, precipitation can include snow, rain, freezing rain, ice pellets or mixtures producing near-zero visibility even at ground level. After the storm, temperatures often fall rapidly, and there are strong, gusty northwest winds and occasional snow squalls. All of these conditions present challenges to the good driver at ground level. At the height of the SCI bridge, conditions would be substantially worse, with the possible exception of reduced blowing snow if it were cleared in a timely manner.

It is certainly possible to construct a bridge which would be safe in any plausible wind and precipitation conditions. To do this would require a structure perhaps 2-2.5m high, with very little air space between the vertical posts. The height would ensure safety for large empty trucks. Such a structure may be very safe in winter but would provide essentially no view of the land or water during the summer. In short, it would be a heavy "tunnel in the air" with minimum tourist appeal. The designs displayed so far by SCI appear to show a solid barrier about 60% to 70% of the height of a typical passenger car — perhaps 1m — with a short guard rail just above. For the gusty conditions on this kind of high bridge with two lanes and a total bridge deck width of 11m, the driving conditions before, during and after winter storms will often be quite challenging and unsafe. In other periods of high winds, or when there is bridge icing without high winds, a great deal of concentration will be required to successfully navigate the bridge.

The net effect of the above conditions is that the proposed SCI bridge will be closed or effectively closed for many hours each winter. The exact number of hours during which the bridge would be shut down cannot be exactly predicted until there is a more detailed drawing of the bridge and a new work up of the wind and precipitation data from Summerside and other nearby stations.

Icing and Safety

No bridge of the length proposed for this project has ever been constructed in a region with frequent freezing rain. In order to maintain safe operations on this bridge in winter, very close monitoring of the bridge deck at several places along its length must be done. Such a system is used on the MacKay Bridge in Halifax. Monitoring of this system, maintenance and repair of the bridge deck, and ice and snow removal all must be properly considered from a financial and environmental point of view. Very little has been said about any of this by the proponents of the bridge.

What materials will be used for the bridge deck? How will the ice be removed? How quickly after the storm will this happen? How will traffic move while the 27 kilometre bridge and approaches are being cleared? How often will it need to be resurfaced? What are the costs of this? How long will the bridge be closed for this operation? How will heavy trucks affect it? Recent resurfacing of Halifax's MacKay Bridge was an engineering and financial nightmare because of errors in materials and a few rogue trucks that used the bridge too soon after the resurfacing. Questions of which vehicles are causing what kind of wear to a bridge are not simple to answer.

In short, the current structure has not been shown to be safe in high winds and in visibility-reducing precipitation. Ice removal will be quite disruptive to traffic unless substantial money is set aside to ensure that it is done quite quickly after precipitation occurs.

The Process Used to Evaluate Climate and Weather Effects

As the reader can deduce from the words above, I have very few kind words for the process which has been used to evaluate climate and weather problems. Let's investigate a few main points regarding this process.

First, there is the question of a bridge versus a tunnel. Everything discussed in this chapter is mitigated if a tunnel were to be constructed rather than a bridge. Yet the tunnel proposals were eliminated by Public Works Canada.

Second, there is the question of local knowledge. In several places it is obvious that the ice experts who worked with the proponents of the bridge were from Calgary. They underestimated the regional climate effects with respect to larger scale climate changes. They dismissed out of hand the thought of regional cooling. They failed to understand the effect of rain/snow changes.

Third, the opponents of the bridge were not given adequate funds to do wholesale vetting of one or more concrete proposal(s), with actual dimensions and materials and costs. The idea of checking out a vague generic design is simply not tenable when so many complex issues are at stake.

Finally, because Public Works Canada was advocating this structure, the line between government as referee and as proponent was blurred. There remained very little independent expertise in Atlantic Canada that had not been "bought" by one or more of the proponents.

In summary, some of the weather effects on the bridge are clear, but too many unanswered questions remain. The one conclusion that can certainly be drawn is that the effects of the bridge on climate will quite clearly prolong winter-like conditions in the region.

Ice in the Strait and its Effects on the Bridge

Erik Banke

The following article is a description of observed ice patterns in the Northumberland Strait, and an analysis of the probable effects of construction of a bridge on those ice patterns.

The Natural Setting

Ice in Northumberland Strait generally drifts from the northwest to the southeast, and is driven by northwest winds and currents. Therefore, it is the ice in the 2400 square kilometre western sector of the Strait that is of concern to bridge builders and fishers and environmentalists. This ice in the western sector normally has to make its way through the narrow passage between Borden, PEI and Jourimain Island, N.B., whereas ice situated east of the narrows normally continues drifting southeast.

Ice in the western sector comes from local ice growth and from ice drifting in from the Gulf of St. Lawrence across the 62 kilometre wide passage between North Point, PEI and Pt. Esquimac, N.B. During winter months, ice typically grounds to a depth of 9m off Borden and to a depth of about 6m off Jourimain Island, N.B. The net result is that the effective width of channel available for the passage of ice is about 11,000m. The Strait is about 12,700m wide at its narrowest point, and can be seen to act as a funnel, with a 62 kilometre wide opening, an 11 kilometre wide spout and boundaries consisting of exposed coastlines and landfast ice in bays and inlets. Because of the reduction in width, ice drifting towards the narrows increases in concentration and ice pressure. This leads to the generation of pressure ridges, rubble fields, and ice pile-ups along exposed coast lines. At the narrows, ice concentrations of $^{8}/_{10}$ to $^{9}/_{10}$ are normal, and $^{10}/_{10}$ concentrations occur regularly. (A $^{10}/_{10}$ concentration indicates that ice covers all of the narrows.)

High concentrations and high ice pressures can lead to blockages of ice in the narrows (Bercha, 1987 and MacLaren Plansearch, 1989). Blockages prevent passage of ice through the narrows, resulting in an increase in ice concentrations in the western sector. Discussions with Marine Atlantic indicate that natural blockages of the narrows have allowed ferries to steam in open water east of the blockage — such a situation occurred in 1976, for example. On another occasion, in 1981, $^{10}/_{10}$ ice concentration and big ice fields jammed the narrows both east and west for two or three weeks — the ice was essentially stationary and ice pressure was observed to vary with the tides.

High winds, dynamic tidal currents (up to 1.3m/sec) and the converging Strait can create severe pressure in the ice, resulting in pressure ridges (up to 35 per linear kilometre) and ice pile-ups along exposed shorelines. In 1976, for example, ice blocks were pushed up over the seawall at Tormentine, and ice pile-ups at Tormentine reef often reach a height of 6m above sealevel. Further west, ice pile-ups to 10m above sealevel were documented at Stonehaven Harbour, N.B., in January 1989. The harbour is situated on the south side of Baie de Chaleur, where ice is exposed to a lesser wind fetch and lesser currents than those observed in the Northumberland Strait. A major portion of the piers, as well as the harbour itself, was buried under ice at the time, and ice was piled up to a height of 10m above the water line (Banke, 1989).

Ice in the Strait and in bays and inlets normally grows to a thickness of 90 cm, and floes in the Strait range from a few metres in diameter to large reconsolidated conglomerate ice floes that are up to 5000m in length. Forward (1959) observed this type of large floe in the Strait and concluded that they must have originated on the west coast of the Magdalen Islands because of the stacks of seal pelts on the ice. The conglomerate floes are generated by the reconsolidation of many smaller floes and rubble fields. Winds from the east cause these large floes to drift west from the Magdalen Islands towards the coast of New Brunswick and then into the Strait. The thickness of reconsolidated conglomerate ice floes ranges from 1.5 to 1.8 times the thickness of undisturbed ice. Ridge keels can, of course, extend much deeper. Fortunately, large conglomerate floes do not appear very often in the Strait. Floe sizes in the Strait have been estimated to be generally below 600m (Bercha, 1989).

Ice Clearing Patterns

Four characteristic break-up patterns have been identified (Forward, 1959). The type 1 pattern is associated with strong northwest winds, and the type 2 pattern with strong north winds. These two patterns of break-up account for 15 of the 19 break-up events investigated, and result in the clearing of ice out of the western sector to melt and drift away in the eastern portion of the Strait. The type 3 ice break-up pattern is dominated by winds from the east, which result in the westward drift of ice into the western sector. This pattern of break-up occurred during 3 of the 19 investigated break-ups. In one of the 19 investigated break-ups, southerly winds resulted in early clearing of ice from the Strait and from the southern marine areas of the Gulf of St. Lawrence.

Date of Ice-out

Bercha (1987) states that the mean date of clearing (or ice-out) occurs around April 23, and Barry et al (1991) in their final report of the ice committee convened in response to the Environmental Assessment Panel (EAP), show that from 1969 to 1988 inclusively, the date of ice-out occurred as early as March 24 and as late as May 28. The great variability of ice-out should be noted when investigating the major concern that delays in ice-out will occur when a bridge is built.

Effects of a Bridge

A generic bridge — with 250m spacing between 10m diameter bridge piers and protection against collisions at navigation piers — will result in increased groundings in shallow water areas, to the 10 or 11m depth contour. The net effect of this increased grounding, and the presence of piers and protective navigation berms (which can also be expected to experience grounding on the berms to the 10 or 11m depth contour), is that a total of about 8900m of segmented channel will be available for the passage of ice. These segments include about 32 spaces at 250m - 10m = 240m, and seven spaces at 250m - 74m = 176m. The presence of a bridge will reduce the channel width available for passage of ice, from 11,000m to about 8,900m (a reduction of 19% from natural conditions), and will divide the channel into discrete segments. This significant reduction in available channel width will undoubtedly result in increased ice concentrations — from the present average 85% to 100% concentration — at the bridge piers.

Ice Jamming Enhancement

It has been acknowledged by Bercha (1987) and MacLaren Plansearch (1989) that naturally-occurring blockages can exist under certain environmental conditions, and Marine Atlantic operators have observed blockages where ice jammed into the narrows allowed ferry traffic to pass through open water to the east. The introduction of a row of bridge piers should result in an increased probability of ice jamming, an increased generation of rubble fields and pressure ridges, and an increased probability of reconsolidation of ice in rubble fields. Considering that the potential for ice jamming is enhanced by quasi-stationary conditions (little or no motion induced by high cross-channel wind events and bridge piers), and by high pressure in the ice, it is not surprising that the jam potential is enhanced.

Jamming of ice at bridge piers can occur in two ways: by smaller floes arching across bridge piers; and by ice floes too large to pass between bridge piers. As soon as a partial jam occurs, the immediate result is that the number of gaps available for the passage of ice is reduced. This has the effect of increasing ice concentrations upstream of the piers, reducing the mobility of ice under pressure and enhancing further jamming.

It is interesting to consider the consequences of a jam at the bridge piers. First, open water will appear on the east side of the jam as ice floes to the east will continue to drift southeast. New ice will form continually during cold winter months and this new ice will also drift towards the southeast. A similar situation exists in Lancaster Sound, N.W.T., where a natural ice edge occurs every year, and new ice grows rapidly in the open water to the east of the ice edge, and then drifts east.

To the west of a jam in the Strait, incoming ice can be expected to submerge at the leading edge and increase the jam thickness when the current speed exceeds a certain critical submergence value. In addition, arriving ice will be compacted into rubble fields and pressure ridges. It is conceivable that the unconsolidated keels of large pressure ridges could reach the seabed in some areas. An ice jam that does not clear rapidly from bridge piers can be expected to grow rapidly in horizontal extent. Suppose the average concentration in the western sector is $7/10$, and the drift rate is 15 kilometres per day, as noted by Black (1959). Given these conditions, the influx of ice from the Gulf of St. Lawrence would cover about 640 square kilometres per day. Ignoring landfast ice areas in bays and inlets, in the western sector, 70% of the 2400 square kilometre area (1680 square kilometres) would be covered by ice when the ice concentration is $7/10$. Open water would occupy 720 square kilometres, and it can be readily deduced that the ice concentration could increase from $7/10$ to $10/10$ in a little more than a day of drift. If the original concentration had been $5/10$, it would have taken about three days to increase the concentration from $5/10$ to $10/10$, but redistribution of ice and rubble fields and pressure ridge generation would probably take an additional day or two. This suggests that in about 4 days, an ice jam could result in 10/10 ice concentration in the western portion of the Strait. In the event that flow of ice from the Gulf of St. Lawrence into the western portion of the Strait is prevented or greatly reduced, it is expected that ice conditions along the New Brunswick coast, north of Pt. Esquimac, and ice conditions along the north coast of PEI will become more severe.

Reconsolidation of rubble ice can result in the growth of 40 cm of solid ice in a matter of 10.5 days when the temperature is -10℃. The life expectancy of ice jams is obviously critical. If clearing mechanisms succeed in destroying partial ice jams quickly, for example by a reversal of currents on the change of the tide, the potential for further ice jamming is greatly reduced and consequently rubble field generation and reconsolidation will likewise be limited. Ice jamming by small floes arching across adjacent bridge piers will probably be short-lived events because the external forces acting to destroy the arch exceed the ability of the arch to resist deformation. Breeching an arch results in the transport of ice downstream. As for ice floes exceeding the span between bridge piers, Bercha (1987) has pointed out that the frequency of single floe jams for a span of 250m is estimated to be: 240 for the Jan. 1 to Jan. 21 period; 240 for the Jan. 22 to Feb. 15 period; 275 for the Feb. 16 to March 15 period; and 160 for the March 16 to April 15 period.

This means that about 10 large floes per day will impact one or several bridge piers each day during the Jan. 22 to April 15 period. During early January, the ice is expected to be so thin that floes will shear past bridge piers.

Ice-out

Normally, a combination of ice drift and melting clears the ice from the Northumberland Strait. Environmental concerns have been raised concerning potential delays of ice-out dates if a bridge is built. There is concern about the effect of such delays on fish catches in the Strait, notably lobster and scallop, which provide a major cash crop.

When an ice jam exists at the end of the ice growth season (usually during the third week of March), the ice east of the bridge will drift away leaving open water. The ice jammed at the bridge piers will then be exposed to three main mechanisms for ice destruction: ablation (melting) of ice at the top surface; enhanced ablation from the underside of ice by the mass flow of warm water under the ice; and wave erosion along ice edges.

When jammed ice increases in temperature, and the strength and thickness decreases, jammed ice will weaken to the point where it will shear past bridge piers and then drift downstream. It can thus be seen that jammed ice must partly decay *in situ* before the ice can be transported into warm water areas. In this respect, the decay of jammed ice differs from the decay of mobile ice. It becomes important, then to estimate the time required for jammed ice to decay sufficiently to shear past bridge piers and drift downstream.

Ice ablation generally commences during the third week in March, in response to the Accumulated Degree Thawing Days. During April a total of 90.8 Degree Thawing Days accumulate, and during May, a total of 283.6 accumulate. About 10 Degree Thawing Days are required to melt 10 cm of ice, so that in April, for example, a total of 90 cm of ice could be ablated from the top surface.

Ablation from the underside of ice is a function of current speed and the temperature difference between ice and the water flowing underneath. For a current speed of 0.2m/sec and a 1 °C temperature difference, a loss of about 10 cm of ice per day has been estimated. At this rate, 120 cm of ice could be ablated from the underside of ice in 12 days. A water temperature of 1 °C is normally reached by Julian day 108 (April 10), and by April 24, the water temperature reaches 2 °C under natural undisturbed conditions. The effects on water temperatures downstream of an ice jam melting *in situ* upstream, would be to decrease the water temperature, and the net ablation on the underside of jammed ice is therefore difficult to determine.

The overall effect of ice jamming on the date of ice-out is difficult to ascertain, but a comparison of ice conditions and dates of ice-out in 1984 and 1985 may cast some light on the subject. In 1984 and 1985, heavy ice existed in the Strait. In 1984, the Strait was virtually free of ice by April 12, but an infusion of heavy ice from the Gulf of St. Lawrence occurred later in April. By early May, the eastern sector was nearly clear of ice, whereas the ice in the western sector drifted around and finally decayed *in situ* by May 25. By contrast, the heavy ice seen on March 20, 1985 had cleared out by April 11. Thus it can be seen that the presence of heavy ice in the western portion of the Strait in late April was instrumental in extending the ice season by more than 30 days (Washburn, Gillis et al, 1989). If this situation is an indication of potential delays in ice-out for ice jammed in the western sector, it is safe to say that, at least during some ice seasons, conditions exist for prolonged delays in ice-out from the western sector of the Strait.

This finding is in sharp contrast to the findings of the final report of the ice committee (Barry et al, 1991).

Conclusions

- Bridge piers will enhance ice groundings in shallow water, out to the 10 -11m depth contour at Jourimain Island and Borden Point.
- A bridge with 10m diameter piers, 250m pier spacing and navigation berms, will reduce the effective channel width of 11,000m (under natural conditions) to a series of 32 discrete 240m passages, and 7 gaps

of about 176m at the navigation piers. The total channel width available for passage of ice is therefore about 8,900m, which represents a 19% reduction from natural conditions existing prior to construction of a bridge with piers spaced as indicated. Different pier spacings and pier diameters will make only small differences in the channel width available for passage of ice.

- Bridge piers can be expected to increase ice concentrations, enhance ice jamming potential to the west, and enhance open water and new ice growth to the east. Another result may be an earlier than normal date of ice-out in the marine area east of a row of bridge piers.
- Ice-out in the western portion of the Strait can be expected to be delayed, depending on the severity and final area of ice jams and environmental conditions.
- If the western sector becomes fully ice covered, and ice from the Gulf of St. Lawrence cannot enter the Strait, the severity of ice along the north coast of PEI can be expected to increase and result in greater concentrations, more pressure and delays in clearing.

References

Banke, E.G., "Ice pile-up at Stonehaven Harbour, New Brunswick," 1989. An ice survey report prepared for the National Research Council of Canada.

Barry, G., T. Carstens, K.R. Croasdale and R. Frederking, "Final report of the ice committee," 1991. Prepared for the Northumberland Strait Crossing Project.

Black, W.A., "Gulf of St. Lawrence ice survey winter 1959," 1959. Geo. Paper No. 23, Dept of Mines and Tech. Surveys, Ottawa.

Bercha, F.G. and Ass. (Ontario) Ltd., "Northumberland Strait Crossing Ice Climate Study," 1987. Final report, and Addendum, 1988. Prepared for Public Works Canada.

Bercha F.G. and Ass. (Alberta) Ltd.," Analysis of photographic data from 1964 to 1965," 1988. A final report prepared for Public Works Canada.

Forward, C.N., "Sea ice conditions in the Northumberland Strait area," 1959. Geo. Paper No. 21. Dept. of Mines and Surveys, Ottawa.

MacLaren Plansearch Ltd., "Study of the sea ice climate of the Northumberland Strait," 1989. A supplementary report prepared for Bedford Institute of Oceanography, Dartmouth, Nova Scotia.

Washburn & Gillis Ass. Ltd., F.G. Bercha And Ass. Ltd. and P. Lane & Ass. Ltd., "Ice/Climate/Fishery Interaction Study," 1989. A report prepared for Delcan-Stone & Webster, Halifax, NS and Northumberland Strait Crossing Project.

Potential Impact of the Proposed Bridge on the Marine Environment and Fisheries of the Southern Gulf of St. Lawrence

Michael J. Dadswell

The Northumberland Strait Crossing Project is a proposal to construct a fixed link for the Northumberland Strait between the provinces of New Brunswick and Prince Edward Island. After years of debate, between 1985 and 1993, it now appears that a bridge crossing may be constructed. This bridge — a high-level concrete structure, spanning the Strait for 13 kilometres — will be one of the longest highway bridges over a sea channel in the world. It will also be the first such structure to be constructed over heavily ice-infested water. The latest plans submitted by Strait Crossing Inc. (SCI) of Calgary, Alberta, are for a cantilevered, box concrete structure with 68 support piers in the water. The marine spans (between piers) will be mostly 250m in length. The height above the sea surface will be 40m, except in the navigational channel, where it will be 64m.

A project of this magnitude, which might cause significant adverse effects on the environment, is required by Canadian legislation to undergo the Environmental Assessment and Review Process (EARP). The proponent of the project must prepare a Generic Initial Environmental Evaluation (GIEE) on the possible biophysical and socio-economic impacts of the project, and have these reviewed by an Environmental Assessment Panel (EAP) appointed by the Minister of the Environment. In the case of the Northumberland Strait crossing proposal, the Panel, which met from April 1989 to August 1990, concluded that the risks to the marine ecosystem of the southern Gulf of St. Lawrence were *unacceptable* (EAP, p. 11), and it recommended that the project not proceed. The decision was based largely on the unresolved issue of whether there would be additional bridge-induced ice formation in the Northumberland Strait, which would cause delayed ice-out in the spring and result in a shift in the ecological balance of the marine ecosystem.

Unfortunately, the Panel's decision was circumvented by the appointment of an ice committee (not a true EAP) to examine the ice question only. The committee submitted the opinion that a bridge would have no appreciable impact on the ice climate of the Northumberland Strait. The ice committee's opinion was accepted by the government, and the project allowed to proceed.

Throughout the deliberations, government organizations such as the Department of Fisheries and Oceans (DFO) and various fishers' groups expressed concern over the potential for environmental impacts and the short- and long-term effects of the bridge on the commercial fishery in the southern Gulf of St. Lawrence. Although the original Panel accepted some of these concerns as valid, the government has chosen to regard potential impacts as minor. Unfortunately, to discover definitively which opinion is correct, it is necessary to perform the experiment (i.e. build the bridge). However, this is not a laboratory experiment, and the effects of a bridge, if undesirable, will be at a high environmental and/or socio-economic cost. Since the information on potential impacts and the reasoning behind why these may occur is scattered throughout a series of letters, summaries and unpublished or difficult-to-obtain reports, the following words are an attempt to explain the concerns and to assess their probability through reference to known and/or hypothesized impacts from other large scale human alterations of the marine environment.

Two concepts that must be understood to appreciate the environmental impact of human alteration on marine ecosystems are the "open" character of the world's oceans, and the long-term cumulative effects of small changes. The first of these concepts follows from the fact that over large regions of the ocean there are few, if any, barriers to the movement of water and organisms. This leads to a great number of interrelationships among physical characteristics of the ocean, such as tides, currents, salinity and water temperature. It also leads to an interdependence among regions for fish stocks. The second concept follows from the fact that long-term cumulative effects of small environmental changes have the potential to change ecosystem structure and to alter the occurence and abundance of organisms in these systems. This concept is conclusively demonstrated by the problems that people in eastern North America experience with acid rain from acid emissions to the west of us.

There are many fish stocks which migrate between oceanographic systems and over considerable distances. The interrelationship between ocean current systems and fisheries was clearly demonstrated by the effects of the Canso Causeway on the lobster stock of Chedabucto Bay and eastern Nova Scotia. Construction of the causeway stopped the flow of water and lobster larvae from the Gulf of St. Lawrence to Chedabucto Bay. Studies have estimated that the reduction in lobster recruitment was around 60%. Lobster landings in Chedabucto Bay declined by 95% between 1955 and 1975 and have remained at low levels into the 1990s. Conservative estimates indicate that the economy of Nova Scotia has suffered a loss of $60-100 million because of these lost catches. The net effect is that the communities along the eastern Nova Scotia coast have suffered extreme economic hardship.

In short, construction of a bridge could have effects over a far larger area than the 14 kilometre distance on each side of the structure that has been identified by GIEE as the area of concern. Rather, the effects of construction could translate into economic problems for fishers in the Southern Gulf of St. Lawrence. None of the studies carried out for the GIEE examined possible far-field effects on the biotic community of the Gulf of St. Lawrence.

On the other hand, cumulative effects were considered to be of concern to the Environmental Assessment Panel during its assessment of the bridge concept. Their importance was played down, however, by the proponents of the project, and possible impact on the marine ecosystem was given only cursory attention in the GIEE. Cumulative effects are similar to compound interest. They do not seem large, but, in time, are capable of major impact. Some readers probably remember that in the mid-1950s, when sea temperatures around the Maritimes increased to about 2°C above long-term means (because of sun-spot activity), there was major die-off of ripe, spring herring in the Northumberland Strait regions because of an epizootic (parasitic) outbreak. This changed the ecological balance between spring-spawning and fall-spawning herring in such a way that, to this day, fall-spawning stocks, which were formerly less abundant, now comprise the greater part of the fishery.

The potential impact of bridge construction in the Northumberland Strait on the marine environment and the fisheries of the southern Gulf of St. Lawrence can be examined from two viewpoints: short-term effects that, with a proper data base and a full knowledge of construction procedures, can probably be predicted with a reasonable degree of probability; and long-term effects, which may be very subtle at first and are extremely difficult to predict with certainty. Short-term effects which are likely to be readily noticeable may not have lasting effects, and will probably operate only within the nearfield region of the bridge (within tens of kilometres). Long-term effects are likely to go unnoticed for a number of years, their impact could be devastating on some sectors of the fishery, and they are likely to occur in both nearfield and farfield regions (hundreds to thousands of kilometres) from the bridge.

Nearfield, short-term effects of bridge construction and presence were fairly adequately addressed by the GIEE's submitted to the panel. These effects include: the loss of benthic habitat because of pier construction; the impact of sedimentation from construction; the impact of construction activity on fish and fishers and the impediment to fishing caused by possible delayed ice-out.

The design submitted by SCI in its Environmental Management Plan (EMP) indicates that a total bottom area of approximately 50,000 square metres (0.05 square kilometres) will be covered by pier construction and protection islands — in other words, approximately 29% of the bottom area under the bridge. The loss of 0.05 square kilometres of benthic habitat, in a region the size of Northumberland Strait, is unlikely to cause changes in the benthic fisheries (scallops, lobsters) that will be distinguishable from among the other factors which cause annual variation in landings (recruitment, fishing effort, weather etc.). The construction of abuttment islands around the four centre navigation channel piers with armour rock may even produce additional lobster habitat (i.e. an artifical reef). Fortunately, the main concentration areas in the local scallop bed are a few kilometres from the crossing site.

The impact of sedimentation from construction is likely to be minimal, as this is one aspect of construction that may be closely monitored. Construction methods are such that the incidence of high sediment loads in the water column should be intermittent, and since the grain size of sediments in the Strait is relatively large, material will settle rapidly. Scallops are quite sensitive to suspended sediments, but since all but the largest ones are mobile, they could move if stressed. Lobsters, to judge from their distribution in the Bay of Fundy (they are abundant in the turbid upper regions — i.e. in Chegnecto Bay), should be little affected by sedimentation. Sedimentation may affect the plants that herring require as spawning substrate and some loss of spawning habitat could occur during construction. Such an impact will be difficult to assess, however, since the GIEE studies on herring spawning sites were too inadequate to conclusively indicate where and with what intensity herring use the local area to spawn. Study of the herring spawning in the Strait is definitely one area of concern that should not have been left solely for the proponents of the bridge and their consultants to study. If DFO truly had the well-being of its clients, the fishers, in mind, the department would have examined this question in detail.

The impact of actual construction activity on fish and fishers will be more difficult to assess or quantify. Little work has been done on the effects of surface and subsurface activity on marine organisms. Also, effects probably differ between species and different modes of behaviour. One can be certain that blasting and pile-driving will have immediate, localized effects (to the benefit of the gull population) and undoubtedly there will be some fish kills. It was interesting to note that in the SCI draft, care will be taken through a Generic Environmental Protection Plan (GEPP) to ensure that no sea mammals or sea birds will be within 500m of the blast site. No such concern is shown for the fish — another example of "out of sight, out of mind." Whether or not SCI will compensate fishers for these kills will no doubt depend on negotiations. Construction activity could have detrimental effects on herring spawning behaviour, but quantification of the effects will be difficult without baseline data. Of course, SCI could demonstrate good environmental and public awareness by suspending construction activity during the herring spawning season — such consideration could reduce the likelihood of construction affecting the long-term viability of the herring stock. Flounders are yet another matter, however. After numerous scuba dives, I have to conclude that they are attracted to places where bottom disturbance is occurring, where they feed on exposed benthic organisms. Prepping a blasting site will attract them. If there is much blasting, there could be a significant negative impact on the flounder population.

The impact of construction activity on fishers will be immediate but localized. The bridge crossing area (a 1 kilometre corridor) will be lost to fishing for periods of time during construction, and fishers will undoubtedly lose gear because of increased surface vessel activity. Mitigation of these effects will probably depend on who complains loudest, and who is the best negotiator (or who has the best lawyer).

The presence of bridge piers and lack of ice-breaking by winter ferries will result in an increased potential for delayed ice-out and earlier ice-up to occur in the Northumberland Strait. Accumulations of ice may occur, depending on the pattern of annual ice break-up. According to the former ice expert for the EAP, ice could extend hundreds of kilometres on either side of the bridge. Obviously, this would disrupt fishing activities in Nova Scotia and New Brunswick as well as PEI. Changes as great as two weeks later than the present average date of ice-out were hypothesized in some of the earlier ice reports. Unfortunately, the GIEE adapted by SCI predicts that the delay in ice-out will not exceed more than two days in 100 years — this prediction is based on their bridge design and hypothetical models. Serious objections to these models were brought forward by the Panel's former ice expert, the DFO and impartial scientific experts from other government departments. They contend that delayed ice-out would affect the beginning of the scallop fishery and cause it to conflict with the lobster fishery in Lobster District 7, downstream from the proposed bridge. Delayed ice-out would also lower spring water temperatures and delay the arrival of the local herring stock. Some of these problems might be moderated by changes in fishing strategies and season. However, if ice-out is delayed and ice-up occurs earlier, it is *certain* that fishers will have a shorter *net* fishing season, which may cause economic problems for the industry in the long term.

The Gulf of St. Lawrence is a region typical of the continental northern temperate zone, with a cool, humid climate, cold winters and hot summers. However, the southern Gulf of St. Lawrence, and the Northumberland Strait in particular, have atypical oceanographic conditions during summer, when compared to the rest of the Canadian Maritimes. Because of shallow depths, restricted tidal circulation and the proximity of large land masses, average summer sea surface temperatures are above 18°C, and the onset of the warm-water period occurs earlier than in the rest of the region, even though the beginning of the ice-free period often occurs later than in the rest of the Maritimes. Thus, populations of organisms occur in the southern Gulf region that are normally not found north of Cape Cod. These include commercially exploitable populations of American oyster (*Crassostrea virginica*) and hard clam (*Mercenaria mercenaria*), and a group of crab species (*Ovalipes ocellatus, Neopanopeus sayi,* and *Rhithopanopeus harrisii*) that are rare or absent elsewhere in Canada and numerous other rare marine organisms. It is interesting to note that in SCI's Environmental Management Plan, arguments document at great length the occurrence of rare plants (those which contribute a major addition to the respective distribution pattern of a species in eastern Canada) in the Cape Jourimain National Wildlife Area, but fail completely to record the presence of rare, *marine* species in the Northumberland Strait. The southern Gulf also supports the fastest growing population of lobsters in Atlantic Canada, and one of the most productive. Sublegal (50-63.5 mm Carapace Length) and small legal lobsters (63.5-80 mm Carapace Length) in this region usually molt twice a year, a situation unique in the Maritimes.

The danger of the long-term cumulative effects of the bridge is probably greatest where it affects the relationship between the rate of spring warming and the total accumulation of thermal units in summer. As long as ice persists in spring, water temperatures will remain low (about 0°C). Studies indicate that a delay of ice-out by two weeks could reduce thermal unit accumulation by about 20-30% on average. Since this situation occurs irregularly under present conditions (it happened only twice in the period 1969-87), it has never had a significant impact on the warm-adapted marine populations. Production losses, higher levels of mortality and lower recruitment during cold years are regained in average or warmer years. However, if the average ice-out period is shifted to later in the spring, permanent cumulative loss of production and/or recruitment could, in time, reduce or eliminate these populations. The proponents' GIEE fails completely to examine the effects of these possible cumulative interactions, the results of which could be felt hundreds of kilometres away from the bridge, could place rare Canadian marine species in danger of extinction, and could threaten the economic viability of important commercial fisheries in New Brunswick, Nova Scotia and PEI.

Oysters, hard clams and lobsters are probably the commercial species that would be most affected by a bridge. The oyster leases of northern PEI depend on seed stock from the highly productive beds around Summerside. These beds are immediately upstream from the bridge, and well within the ice field region that will be generated if the bridge causes ice to jam in the Strait. Loss of productivity in these beds would affect the viability of the entire PEI oyster industry. Similarly, the productivity of hard clam beds would be reduced by a general lowering of heat accumulation in summer.

There are two possible long-term effects of delayed ice-out that could have a significant impact on the lobster stocks of the southern Gulf of St. Lawrence: reduction of growth (i.e. reduction of productivity) causing average landings to decline, and the cumulative effect downstream of reduced recruitment in lobster stocks in the eastern Gulf. Lobsters require a specific amount of heat accumulation (thermal units = degree days where 105 days x 10°C = 1050 degree days) above a base temperature of 5°C in order to molt (or grow). A 10-20% decrease in heat accumulation because of delayed ice-out would cause a decrease in the overall rate of molting each year in the Northumberland Strait. The immediate effect would be a loss each year in Lobster District 8 of those lobsters that would otherwise have molted a second time, reached legal size and entered the fishery. The long-term effect of lower growth rate would be an overall reduction of landings. In effect, it would be as though the Northumberland Strait had been moved further north. It became evident to members of the Panel during their deliberations that "the risk of loss of production of lobster resulting from delayed ice-out caused by the presence of a bridge is unacceptable" (p. 13, EAP Report).

During the deliberations of the Panel, members requested that the proponents of the bridge examine the effect of a reduced molting rate on the productivity of the local lobster stock. The opinion of technical experts for the proponents was that landings would decrease in the short term, but in the long term overall landings would increase because of reduced fishing mortality on the lobsters that did not molt a second time in a given year. Unfortunately, the experts consulted by the proponents failed to consider that overall *average* growth rate would be reduced by an *average* lower heat accumulation. The model they used to estimate landings was flawed because it did not recalibrate with a different growth equation that would account for the lower temperatures caused by delayed ice-out.

The cumulative effects downstream of reduced recruitment in lobster stock in the eastern Gulf of St. Lawrence may become even more significant than reduced growth rates. Organisms such as lobsters, with reproductive cycles that include a pelagic, larval stage (i.e. babies drift in the currents), fit themselves to the prevailing environmental conditions through their reproductive behaviour and natural selection. For the lobster, this includes upstream migration behaviour in mature individuals which counteracts the average downstream drift of the larvae. Organisms evolve in order to exploit the prevailing situation and develop what are known as "recruitment cells." The recruitment cell encompasses all attributes of the organism's biology (reproductive type, nursery needs, adult feeding, etc.) and the oceanographic system in which its population exists. The recruitment cell of the Northumberland Strait lobster stock extends from just west of the bridge crossing area, downstream to northern Cape Breton, and once included the eastern shore of Nova Scotia. It was the death of the eastern Nova Scotia lobster stock through closure of the Canso Causeway that caused the 95% collapse of this lobster fishery between 1965 and 1975.

Scientific studies conducted in the past have hypothesized that the lobster landings (and therefore the lobster population) of the eastern Northumberland Strait were a unit stock. If this is true, it should be possible to predict that there will be a region of accumulation of large lobsters in the upstream end of the recruitment cell. An examination of the scientific literature revealed just such a concentration in the Hillsborough Bay area downstream from the proposed bridge. The presence of large, egg-bearing females in this region was confirmed by discussions with the local fishers during the Panel's deliberations.

Lobster reproduction, like growth, is affected by temperature. A certain number of degree days of heat accumulation above a base temperature of 3.5°C is required for the eggs to develop. Since female lobsters carry their eggs for 11 months (from August of one year to June-July of the next) before they hatch, development rates will be affected not only by delayed ice-out in the spring, but also by early ice-up in the fall. Lobster larval growth rates are also affected by reduced temperature, which increases the larval pelagic stage by about 5-10 days for every decrease of 1°C in the average sea temperature. A reduction in average sea temperature in the Northumberland Strait caused by later mean ice-out time would lead to later times for egg hatching, increased drift periods for larvae and increased levels of mortality while drifting because of the longer drift time. The cumulative long-term effect of these interactions would reduce turnover rate of the population, decrease productivity, and decrease landings on all shores of the southeastern Gulf of St. Lawrence in a region extending from the bridge to northern Cape Breton or beyond.

The proposed bridge to PEI is another example of the tendency of governments to force through the most expedient solution to a problem. The Panel expressed a desire for the government to examine other options to the fixed link problem, such as improved ferry service or a tunnel. A tunnel under the Northumberland Strait would, on all accounts, have virtually no effect on the ocean environment or the fisheries, just as a bridge built across the Canso Strait would never have had the effect on Maritime fishery resources that the Canso Causeway has had. It seems that governments, while paying lip service to sustainable development, are doing everything in their power to destroy the very basis upon which the Maritimes could establish sustainable development — its fisheries. The fisheries, not tourism or industry, is (or was) the greatest earner of cash in the Maritimes. To see it destroyed piece by piece, and replaced by some other hypothetical cash source, is difficult to bear (think of the heavy water plant at Canso). These actions seem to comprise one more step towards the urbanization of the world — urbanization that leads to a loss of diversity and productivity on land and in the sea.

The Maritimes, even now with all its problems, has one of the best fisheries left in the world, and every effort should be made to enhance and preserve it. Fishers, and the people who depend on them to keep small Maritime communities alive, should do everything they can through the political process to make the government fisheries departments (and particularly DFO) become better advocates for the fishery. If they were, we might not continue to see the Maritimes fisheries reduced by other vested interests.

Island Society

"It would be the 1990s equivalent of the CPR, a Disney World type of project, one of the modern wonders of the world that people would travel to, just to see it."

Tom McMillan,
Globe and Mail, 25 June 1985.

When Public Works Canada (PWC) selected its five evaluation criteria, it also selected the battlefield upon which the opposition would do battle. PWC has generally been successful in confining the debate to just those issues — economic, financial and environmental. The other two criteria, "proven management capability" and "technical soundness," never became issues. Public questioning of technical capability has remained mute largely because, up to now, details of the bridge design have not been made public. There is, however, a growing body of concern over construction of a bridge which is at the very farthest reaches of engineering know-how, possibly even surpassing such knowledge. Engineering, scientific and trade journals regularly question the wisdom of new types of bridge construction. A few years ago a major American contractor closed down its bridge division, explaining that one reason is that "high-tech bridges are still an R&D proposition." Clearly this is an area in need of more investigation.

In this chapter of the book we step outside PWC's circumscribed battleground to explore beyond the environment and economics. We ask: "What else are we jeopardizing if we accept this bridge project?" Although PWC's overhasty reply is "Nothing, my dear," others feel that issues which have traditionally troubled Islanders are present in a new and more troublesome dimension.

With the construction of a bridge, long-held concerns for the land and a struggling resource-based economy are resurfacing. New alarms are being raised about a project which will be privately owned, encourage unchecked tourism growth, and threaten the use and ownership of the soil into which generations of Islanders have sunk very deep roots, and which they have fertilized with their sweat and blood.

Finally, what has been the media's response to these concerns? Have they — the supposed independant voice of our society — bothered to leave PWC's battlefield and examine *these* crucial issues, as well as those of economics and environment, from any perspective other than that of the Chamber of Commerce? Have they raised a question about the connection between this bridge project and other neo-conservative policies of the federal government? About the privatization of our transportation infrastructure? Or about the results of unfettered tourism development? Unfortunately, no. More accurately, they have "covered" the issue by press release and sound bite — hardly the actions of a dynamic, critical media.

Recalling the Unanswered Question

Kevin J. Arsenault

There has not been a scientifically credible environmental and social assessment of the fixed link now being proposed, but the federal government isn't about to admit that fact. It is clearly not interested in acknowledging deficiencies in any of its environmental impact studies. Is this also the case for the PEI provincial government?

The former Premier of PEI, Joe Ghiz, promised Islanders that his government would not give the go-ahead to the bridge project without first ensuring that it would not cause any unacceptable, negative environmental and socio-economic effects. This promise has been broken. The argument in this article is simple and straightforward: Because there has been no credible environmental and socio-economic impact assessment of the effects of the bridge project, the provincial government has not lived up to its commitment to the people. It should, therefore, serve immediate notice to the federal government that the bridge project will be put on hold until the necessary studies are commissioned, and the results of those studies are evaluated by Island residents in public hearings.

The Federal Response to the EAP Ruling

Federal politicians and administrative bureaucrats were not long in responding to the "don't build a bridge" recommendation of the Environmental Assessment Panel (EAP). The minister for Public Works Canada (PWC), Elmer MacKay, confidently informed Island residents that the federal government had no intention of giving up the project, and was seemingly proud of the fact that he wasn't legally bound to abide by the recommendation of the EAP. Perhaps this was true enough. But was there no point to such an intensive and expensive exercise in democracy? After all, wasn't the decision made by a legitimate and impartial body invested only with the humble duty and simple mandate of noting what dozens of expert witnesses, consultants, organizations and individuals reported orally and documented in written submissions? Not to worry about that, came the answer, PWC would commission new studies.

The politically cynical Islander no doubt predicted this answer from the federal government. Still, it is significant that the federal government's response to the fixed link Panel ruling represents the first time that the federal government has decided not to abide by a "no" ruling of a federally-appointed EAP. Just as troublesome as the initial decision to proceed with the project, notwithstanding the "no" ruling of the EAP, is the manner in which PWC has proceeded since: The federal government has given the go-ahead before properly addressing and answering those key questions raised by the EAP. So why has the province not stood up and raised this matter as a serious problem? This is a good question to think more about, but let's first review the *federal* response and follow-up to the "no" ruling of the EAP.

PWC's response to the EAP ruling was to commission new studies on one issue. It challenged the Panel's findings on a single issue — namely the Panel's concern over the negative effect of the bridge on the Northumberland Strait fishery. This negative effect would be caused by a delay in spring ice-out, which would in turn be due to the support pillars of the bridge keeping the ice in the Strait.

During the time when these studies were being done, the entire fixed link issue more or less drifted out of the focus of the media and the public. Then the inevitable happened. The Minister of Public Works was back in the public eye with good news: His "hunch" about the ice-out concerns not posing an insurmountable barrier was borne out by the tireless studies and ingenious ice-out models developed by new and improved government scientists.

The federal government led the media down the garden path with its exclusive focus on ice-out. Islanders heard nothing but technical scientific jargon about ice for days on end. These reports and interviews with numerous ice experts left many people with the false notion that the problem with ice was the *only* environmental problem raised by the EAP, and therefore, the only obstacle standing in the way of a decision to give the bridge project a green light.

Of course, the EAP's recommendation not to proceed was based on a great deal more than its concern with the ice-out delay. The decision whether or not to build a bridge did not hinge solely on the question of whether Public Works Canada had succeeded in eliminating this one "final" barrier. A whole host of other unanswered questions were lost in the ice-out shuffle. No doubt the EAP would have jumped back into the debate to clear up basic misconceptions — if it had not already been "disbanded" by the federal government. Still, at least two members of the Panel spoke out publicly to remind Islanders that they were being hoodwinked by the federal government with the notion that the EAP's recommendation not to proceed was hooked to one technical issue of "ice-out delay."

This is how the federal minister and bureaucrats within PWC used the media to effect a masterful sleight of hand, and succeeded in obscuring the other condemning findings of the EAP. They succeeded in reducing a lively public debate on the effects of a fixed link on both the Island environment and lifestyle of Islanders to a technical and largely speculative discussion about ice.

We must recall some of those other concerns raised by the EAP if we are to demystify the issue and realize why the decision to build a fixed link has been made prematurely. We can begin by recalling the critical analysis of PWC's own impact studies, offered by Dr. Philip Byer, technical expert to the EAP. Although the media clearly focused on his comments concerning the risk of environmental damage from bridge accidents such as chemical spills — and, of course, the risks to the fishery from too much ice — the most serious deficiency Byer detected in PWC's impact studies was the failure to provide an assessment of socio-economic impacts, especially negative ones.

The Art of Assessing Socio-Economic Impacts

Assessing socio-economic impacts involves far more than a standard cost-benefit analysis. Impact assessments of mega-projects must consider and evaluate impacts on many "intangibles," to which no dollar sign can be easily attached. That is not to say, however, that such impacts are "value-neutral."

Environmental impact studies must reflect concern not only for the effects of development on the natural habitat, but also on the economic, political and cultural lifestyle of people. A credible assessment of the socio-economic impacts of a fixed link must, therefore, first determine what the majority of Island residents truly value and want for their future.

The cost-benefit analysis presented by Public Works Canada fails miserably in this regard. It included only a few items such as capital expenditures, operating and maintenance costs, time-saving costs, taxes, and vehicle operating costs — only items that fit neatly into a statistical computer software program. What couldn't be given a dollar sign was ignored. No consultations, studies or surveys were commissioned to obtain information on the long-term development aspirations of PEI residents.

Apart from numerous particular errors and inconsistencies, the most serious flaw in PWC's impact studies is with the general, positive assumption that adverse effects — no matter how bad they might be — can be mitigated. Byer also noted an obvious tendency within PWC to downplay or avoid altogether what *could* plausibly happen. The most basic deficiency in this regard concerns the crucially important issue of the rate of increased traffic flow that can be expected to result from a fixed link. But this deficiency brings us back to the unanswered question which the EAP directed not at the federal, but rather the provincial, government.

Recalling the Unanswered Question

The EAP had determined that a key stumbling block standing in the way of a credible environmental and socio-economic assessment was the need for clarification from the PEI government on its tourism policy. It noted that before it is possible to assess the negative environmental and socio-economic impacts from increased traffic across a fixed link, it is first necessary to know roughly how many more people a fixed link would bring to the Island, and whether that number would exceed an optimal carrying capacity. Neither of these things had been determined. The EAP stated in its final report that "... there are both environmental and sociological limits to consider in determining desired tourism growth rates and these were not adequately dealt with in [Public Works Canada's] Bridge Concept Assessment." The EAP had presented this concern to PWC by requesting that the federal government examine "... worst plausible scenarios for impacts due to increased vehicle traffic."

PWC responded by suggesting that only the province could answer this question, because the PEI government has jurisdiction over tourism policy. The EAP accepted this response, and subsequently recommended in their final report that "... an optimal carrying capacity for tourism be identified by the Government of PEI" which would "... reflect the aims and aspirations of PEI residents."

No studies have yet been commissioned by the province to provide an acceptable answer to this key question. And, like the federal government, the provincial government has given the go-ahead to the project, despite not providing the information requested by the EAP. Let us examine more carefully the implications of the province's decision not to respond to the EAP question.

Additional Reflections on the Unanswered Question

How many additional people would a fixed link bring to PEI for, let's say, each year of the projected 100-year life span of the bridge? This key question must be answered before it is even possible to measure and assess the socio-economic impacts of a fixed link. Only after the appropriate surveys are done can the impacts of a bridge on the natural environment, infrastructure, economy and culture of PEI be reasonably predicted and assessed. PWC based their entire impact assessment on the assumption that there will be no appreciable increase in commercial traffic as a result of a fixed crossing.

Dr. Byer also found problems with PWC's worst case scenario projections for traffic increase: "The use of a one-time 50% increase in traffic in 1995 [by PWC] as the worst plausible scenario seems reasonable, though one could imagine that in the long term, e.g. 20 years, the bridge together with changes in tourism opportunities could result in an even much greater increase than would happen without the fixed link. In addition there is no statement of the likelihood of this scenario, as requested in the Panel's question."

PWC's refusal to study the likelihood of unacceptably high levels of traffic to the Island represents a fundamental flaw in the government's assessment method — an error which invalidates just about everything else said concerning socio-economic impacts. Again, without knowing how many people might visit the Island during tourist season, how could impacts be measured? Although PWC did commission a subsequent study into projected traffic flows, which was undertaken by Smith Green & Associates, the conclusion drawn by the consulting firm was: "Survey results have indicated the likelihood of increased tourist visits if a bridge crossing were to be constructed. There is insufficient data to assess the possible impacts of a bridge on long-distance markets, such as Quebec, Ontario, and New England, and this needs to be further explored." This question has not been further explored.

Determining the extent to which a fixed link would increase visitor traffic is a formidable task and is certainly nothing like an exact science. Considering changes in traffic flow from other similar projects provides, however, one key indicator to what could be expected on PEI. In March of 1988, a Canadian delegation visited Europe to observe and study other link projects. Although the delegation's final report offered only observations and limited information, it did attempt to assess the environmental and socio-economic impacts of several fixed links. The task force studied four European projects. Of these, only the Öland Bridge and various smaller Norwegian links and tunnels were completed and operating. The delegations' final report offers one quite startling tidbit of information regarding increases in tourist visitor rates for both completed fixed links they studied: "One thing common to both the completed projects visited was that initial traffic increases as a result of the fixed links being in place was substantially underestimated." What does "substantially underestimated" mean?

The 6 kilometre bridge linking the Island of Öland to Sweden began operation in 1972. Before the bridge was built, approximately 500,000 vehicles made the crossing by ferry each year. In the first year of bridge operation, the traffic increased by 500%. The traffic rate peaked at 4.2 million in 1979, and has since stabilized at around 3.9 million annually, a rate nearly 700% higher than that under the previous ferry service.

These significant increases in traffic suggest that earlier projected assessments clearly failed to factor important variables into the calculations. The delegation also noted that during the tourist season, Öland is populated with 5 times as many visitors as permanent residents. This is roughly the proportion of tourists to residents on PEI *before a fixed link*. What will it be after the link is built? And what about Byer's concern over too many tourists flooding onto PEI in, say, 30 years after the bridge is built?

"Optimal Carrying Capacity" & Provincial Tourism Strategy

The majority of Islanders are clearly opposed to a tourism strategy based on uncontrolled, continuous growth. Common sense tells us that no amount of growth in this sector of the economy will answer the need for integrated and sustainable social, economic and cultural development. Sociologists and social planners have carefully examined tourism impact studies and have concluded that every community or region has a point beyond which more tourists are too many — every ecosystem has a threshold beyond which negative impacts reduce the sustainability of the environment and the quality of social life.

In an extensively researched book, *Tourism: the Good, the Bad and the Ugly* (Lincoln, Nebraska: Century Three Press, 1979) the authors stress the importance of identifying a saturation point for tourism and then designing a clearly defined tourism strategy. This will ensure that proper limits and controls are readily available so that tourist growth does not exceed the point of saturation. To achieve such control, the authors suggest that: "In enforcing limits, locals should not be afraid to create artificial tourism constraints such as stopping airport expansion if that is necessary to maintain the liveability of their community and the beauty of their environment." Controls on incoming transportation are clearly the most effective means of ensuring that the rate of visitors does not escalate beyond what the local people desire or the environment can sustain. Wisdom in planning comes with preventing too large a flow, not in preparing to deal with an overflow after it comes.

Given the possibility of significant traffic increase from a bridge across the Strait, we need to know how many additional visitors a bridge to PEI will bring, *before* (not after) a link is built. Knowing that government officials and planners "substantially underestimated" projected traffic flow with similar mega-link projects — such as the Oland bridge — Islanders have reason to be suspicious when bureaucrats from Ottawa tell them not to worry.

At what point would Islanders cut off traffic flow? There have been no studies by either the federal or provincial government to provide an up-to-date answer to this fundamental question. The answer to this question could, in itself, make a fixed link completely unacceptable, given the likelihood of significant traffic increase in coming years. The rate of urban sprawl in the New England states and other North American cities is sure to bring increased traffic to PEI as the condition of the general environment continues to worsen and city-dwellers seek some solace in the relatively pristine environment of places like PEI.

We're not completely in the dark on this question of what Islanders think about increasing tourism on the Island. Relevant studies from Island surveys do exist, although they quickly became forbidden history — i.e. they were neither cited nor included in the superficial tourism inflow studies commissioned by Public Works Canada.

Results from 1974-76 Tourism Impact Studies

The 15-year joint Federal-Provincial Comprehensive Development Plan had, as one of its key objectives, the expansion of the tourism industry on PEI. As imported urbanite planners revealed more and more blueprints for yacht clubs, resorts and golf courses, Islanders became increasing distressed with the direction Alex Campbell's Liberal government was taking with the sudden and rapid development of tourism.

Under considerable pressure from Islanders, Premier Campbell commissioned an independent study in 1974 to determine the impact of tourism on PEI. In May of 1974, ABT Associates Inc. was granted an $84,000 contract to undertake this study. Two years later, the firm made their findings and recommendations public. Through extensive public opinion surveys, ABT Associates Inc. determined that 76% of Islanders did not want any further increase in the number of tourists coming to PEI, adding that the government should not be spending money to promote the tourist industry. The consultants themselves concluded that a doubling of the number of tourists during the following 8 to 10 years would have serious negative impacts on the environment and Island life. In a meeting with the provincial cabinet, the consultants stated that the Island had not yet reached the *environmental* limits but had already reached the *sociological* limit. This is an interesting conclusion, considering that the sociological limit, reached back in 1976, has been virtually ignored in the fixed link environmental assessment.

Premier Alex Campbell — who was no doubt taken aback by the consultants' findings — responded cautiously by saying that the Island could probably hold a few more tourists, but acknowledged that it was clear that Islanders did not want more growth in tourism. Even the Tourism Industry Association of PEI agreed that doubling growth was unacceptable, recognizing as well the need to implement a strategy to control growth (*Guardian*, Feb. 14, 1976).

Are these survey results still relevant? The Island certainly hasn't grown any bigger. With increased public awareness of the need for more environmental protection of PEI's limited land and water resources, one can only conclude that they are more relevant than ever.

Where does the current liberal successor to Alex Campbell stand on this critically important issue regarding growth in tourism? Premier Callbeck has inherited the tourism policy of her predecessor, a strategy futuristically dubbed, "Tourism 2000." This blueprint for continuous growth in tourism does not even mention the need to control tourist growth, nor the need to establish a saturation point or "optimal carrying capacity" for tourist growth on the Island.

According to a 1988 government document, "A Statement on Tourist Development," officials predict a steady 3% growth rate in the number of annual visitors. The aim is to achieve a goal of over one million visitors annually by the year 2,000 ("A Statement on Tourism Development," October, 1988). That's roughly 9 times the number of residents of PEI, and at least a doubling of the number of tourists visiting in the early 1970s, something everyone — including tourist operators — recognized then as unacceptable.

PEI's industry minister, Robert Morrisey, suggested in his submission to the Environmental Assessment Panel that the Panel keep their focus on the "environment," and leave socio-economic concerns to the "government." People are, however, an integral part of the environment, and impacts on people must be given due consideration in a credible impact assessment of proposed mega-projects. It would be foolhardy to leave the concern over socio-economic impacts to the provincial government. The PEI government is currently pursuing an irresponsible tourism strategy which does not even establish an optimal carrying capacity or "cut-off" for tourist growth. Furthermore, the government is ignoring the cultural and social aspirations of Islanders — at least those expressed in the most recent and reliable comprehensive survey of Island opinion on the matter, namely the ABT tourism impact study. Moreover, the PEI government has failed to answer the EAP's key question.

Conclusion

The point of conducting environmental assessment reviews and impact studies is to assess the likelihood and extent of negative impacts *before* they happen, so that inappropriate development projects can be scrapped. The Environmental Assessment Panel has concluded that the bridge should be scrapped. The "we'll cross that bridge when we come to it," approach of the provincial government suggests that the bridge should come first, then the impacts will be assessed after it is too late to do anything about them. That's unacceptable.

The provincial government cannot honestly tell Islanders that a bridge across the Strait will not result in unacceptable, adverse environmental and socio-economic impacts. The most responsible course of action available to the government is to heed the primary recommendation of the EAP: not to proceed with a bridge. Premier Callbeck should serve immediate notice to the federal government that an answer to the key question presented to it by the EAP has not yet been answered, and her government is, therefore, withdrawing support for the project until such time as the appropriate surveys and studies are commissioned. Studies are needed to determine what would constitute an acceptable saturation point for Island tourism. Studies must also be commissioned to determine whether or not a fixed link is likely to bring numbers of visitors that exceed the saturation point.

Let's hope that the province doesn't try to argue that PWC has "satisfied" the concerns raised by the EAP. They haven't. Most of the important unsettled issues have not even been studied — a fact which any member of the EAP would happily confirm. And besides, the question which the EAP asked the pronvince to address is the question that has collected the most dust on the shelf.

Islanders might expect to be denied due process from the current federal government, but it is more than Islanders should be asked to accept from their own provincial government. Surely the Premier would agree that Islanders still have the right to have socio-economic impacts studied and evaluated using an ethical (as well as economic) cost-benefit analysis of impacts. Will the new Premier honour the promise made by her predecessor, put the project on hold, and commission the required studies? Let's hope this doesn't become yet another unanswered fixed link question.

Whose Fixed Link? Control of the Land on PEI

Reg Phelan

Whose fixed link will this be? And what effect will the link have on the Island? These are two important and interconnected questions that we as Islanders must answer for ourselves. They will have an impact on a very important aspect of PEI — our land: the land we as Islanders have fought for and cultivated for generations, the land upon which we live and work, and the land that, in turn, sustains us in a variety of ways. In this article I will draw upon recent developments in agriculture on the Island — developments that are the result of free trade — and show how the proposed bridge will become part of a process that has been set in motion by free trade. This process will lead to the inevitable exhaustion of the land, decreasing economic diversity and an increasing reliance upon one industry, tourism.

We have never received a direct, factual response to the first inquiry. Paul Giannelia of Strait Crossing Inc. (SCI), the consortium that has contracted to build the bridge, has not revealed the identities of the new Landlords of the Strait, saying only that they are "global" investors. And so we must ask: Why do transnational companies want to spend about $840 million on a fixed link to PEI?

To find an answer to these inquiries, it is necessary to acknowledge PEI's history of involvement with private developers, a history that informs our present attitudes and perceptions about a major private undertaking such as the fixed link. We must also take a hard look at what proponents of the fixed link say the mega-project will do for PEI. Proponents of the link, who call themselves "Islanders for a Better Tomorrow," include, among others, local Chambers of Commerce, the Tourism Industry Association of PEI and the PEI Real Estate Association, and construction interests. They believe that a fixed link is necessary if PEI is to be competitive in the global market. Are these promises of growth a possibility and are they sustainable? What effect will these proposals have on our major resources of land and water? An investigation of these questions will enable us to explore the options for building a sustainable future.

The would-be Landlords of the Strait are not unique in their attraction to PEI. The Island has garnered attention from foreign interests for centuries. Jacques Cartier described the Island in these words: "All the land is low and the most beautiful it is possible to see, and full of beautiful trees and meadows. This is the land of the best temperature." It soon became clear, however, that agriculture was only to exist in early settlements if it supported the mercantile interest of the fur trade. Land grants were issued to supply surplus food and support the trading efforts of fortresses like Louisbourg. Communities expanded and grew, and the French settlers refused to abide by the dictates of the British feudal system, but such resistance exacted a harsh penalty — deportation.

After the deportation, the British divided the Island into 67 lots and, using a lottery system, gave the Island to those in line for patronage from the British crown. Most of these landowners remained in England, using land agents and lawyers in Charlottetown to manage their affairs. The land agents were resented by tenant farmers, and resistance grew against them. For the next 100 years, Island history is largely dominated by this resistance and a call for Escheat — a return of the land to the Crown, and redistribution of land to those who actually worked it. The ideological feelings of the era were articulated well by William Cooper, the leader of the Tenants (Escheat) Party, who, in speaking about the absentee landlords to the Assembly in 1832, said: "Did they use or occupy the lands themselves? No. Then why did they wish to hold the lands they could not use or occupy? It was not the lands they wanted, but by holding the land, to have claim on the labour of their fellow subjects who have equal rights with themselves ... " Cooper and other tenants argued that the unrestricted accumulation of private property was opposed to the well-being of society. They argued that the role of the state was the protection of the people and their liberties, not the protection of property at the expense of liberty.

Given our history, people should not be surprised that Islanders are asking similar questions today: Who are the landlords of the mega-project that will be our main transportation link to the rest of the world? Why do they wish to own and control such a vital link? By holding such a link, how much control will they exercise over what happens and is produced here? Will the direction of local Island production, labour and land resources change?

Groups promoting the fixed link, such as Islanders For a Better Tomorrow, are united in promising growth in the Island economy if the mega-project goes ahead. But from where is this increased growth to come? There may indeed be short term growth in the construction industry that will benefit certain areas of the Island economy in the years during which the bridge will be under construction. Their argument for the prediction of continued growth after the construction stage, however, depends upon the restructuring associated with free trade. This argument, of course, assumes a number of conditions: that the Canada-U.S. Free Trade Agreement and the North American Free Trade Agreement (NAFTA) will be in existence after the next federal election; and that the predicted economic activity will actually occur if the changes associated with free trade continue to take place. Of course, it is true that changes resulting from free trade can happen with or without a fixed link. It is important to understand, however, that the private forces associated with the fixed link will powerfully influence the *extent* and *speed* of change, particularly in the tourism industry. But what *kind* of growth will this be? And what impact will free trade and the fixed link bring to our vital resource industries?

The answer for fishers is clear: There is no possibility of short-term growth in the fishery, due to the severe depletion of resources. As for agriculture, the free trade deal — and the privatization and deregulation moves leading up to it — have already had a detrimental effect. Let's look more closely at agriculture, which is so tied to issues of control of the land.

Agriculture on PEI can be divided into three areas:

1. The first area is small fruits and vegetables, with potato farming comprising, by far, the largest part. In this area there is no marketing structure or stabilization program. Prices are determined by a manipulated marketplace. If the market forces know that there is 5%

overproduction in this area, prices can drop by 500%, as we have seen happen this past year.

2. The second area is dairy farming. On PEI, this is comprised largely of cheese-making, and is in a supply-management market situation. Returns are associated with the cost of production and the supply is limited to what is consumed in Canada.

3. The third area is comprised of hogs and beef. The price of livestock is decided by a manipulated market, similar to the one for potatoes.

Before free trade, a stabilization program for hogs provided 90% of the cost of production in order to ease producers over the low times in the marketplace. The stabilization program has been gutted due to the implementation of free trade legislation. As a result of the deregulation of transportation, grain costs are now higher here than in other parts of the country, and many producers have gone out of business — in short, hog farming is in rapid decline. The free trade legislation has also allowed the importation of processed dairy products, and with this the supply marketing system in dairy is threatened. If these threats are realized, numerous dairy producers will be out of business and production will drop.

Free trade has left all the agricultural eggs in the potato basket, so to speak. PEI produces a high yield of decent quality potatoes that have earned a reputation in the market. This, along with our ability to hold a greater portion of the high value seed market, has put us in a supposedly competitive position — if such a thing can really exist in a manipulated global market. Because we are an island, we have been able to monitor diseases and insure a high quality seed. Potato production in the past few years has experienced a large increase . In fact, this year's production was so high, there was a major scramble to find enough storage room. There are not enough trucks in the PEI system to move the present quantity of potatoes to market in an orderly fashion, and potatoes are considered a back haul to imported consumer goods. If the Island were to produce a volume of potatoes greater than the volume of consumer goods, the cost of transporting the extra volume will more than double. A fixed link may reduce the cost associated with transport time, but will never come close to meeting the doubled cost for extra volume. And this is where problems associated with the fixed link, in combination with the results of free trade, become inseparable and insidious.

With the free trade induced devastation of the hog industry and the threat to the supply-management system in the dairy industry, farmers in those two areas will turn to the only remaining agricultural sector — to potato farming. The increased production of potatoes will have a devastating effect on the land base of PEI. The soils of the Island are particularly susceptible to deterioration when a lot of chemicals are used in farming. The quanitity of organic matter in the soil will drop rapidly, and the soil will lose the fibre and texture that holds it together. Dairy farming, the stable force in PEI agriculture for years, grew the forages and provided the manures to build up the organic matter in our soils. Potato growers actively search out land from dairy and livestock farmers upon which to grow potato crops because they know the quality and yield of their crop will be much better on this land.

After a rain storm, many rivers run red with Island topsoil that has washed into them. This erosion is largely the result of practises used in potato production. Some springs, when erosion is severe, bulldozers are required to level the gulleys in the fields that are left. Losses have been measured from 15 to 30 tons per acre in fields with slopes as small as 12%. Some fields have had losses of 65 tons per acre. Longer rotations, division terraces and cover crops help to contain some of the problem. The serious problem remains, however, that the land will no longer be able to sustain the increasing production forced upon it. Lower crop yields and problems with quality will show up in the field and later in storage, and the weakened plant will be more susceptible to insect attack and less resistant to disease. Even now, insects feeding on weakened plants are becoming resistant to increased doses of insecticides — recently, the frequent use of insecticides has not had the desired effect on the Colorado potato beetles.

Farmers are increasingly being made aware of the environmental limits of the current patterns of production. The causes of these limits are complex and deeply rooted in the system of industrial agriculture. And with the push towards free trade, farming is in a state of uncertainty and change. Farmers are having to cope with problems for which there are often no solutions. At the same time, agribusiness regards the farmer as the bottom-rung supplier of cheap raw material for industry. Their attitude is that output should be increased by more input — mostly petroleum-based products — and when the soil becomes mined and this approach doesn't work any longer, production can be shifted to other places where this kind of exploitation can occur — places like Chile that have large foreign debts to feed through the export of cheap food.

It is clear that the highly capitalized industrial model has depleted the fisheries and is eroding our agricultural base. Tourism is the one area left to take up the slack. The number of people visiting the Island has remained stable in the past few years, and has reached an optimum level in most Islanders' minds. A greater number of tourists can lead to the disruption of Islanders' lives, particularly those who make a living in other ways than through tourism, and and can have a detrimental effect on the environment, land and culture.

This is the precisely the point at which decisions about the fixed link become vitally important. Allowing a private consortium to own the fixed link could drastically change the type of economic activity —

particularly tourism activity — practised on PEI. The Landlords of the Strait will have a strong vested interest in moving as many vehicles as possible over the link. This is where they stand to make money, which is their reason for building and owning the structure in the first place. Once the break-even point is reached (with the help of federal subsidies) the bridge is a money-making machine. Every additional car that can be enticed to travel over it represents near total profit. What an incentive for the Landlords to team up with the PEI Real Estate and Tourism Associations and literally sell PEI!

If the present trends in agriculture prevail, financially distressed farmers, whose soils will be exploited by monoculture farming practices and therefore unable to sustain the increasing production required of them, and will not have many choices and will be vulnerable to exploitation.

In a brief to the Environmental Assessment Panel (EAP), in March 1990, the PEI Real Estate Association says: "We anticipate greater investment opportunities will open up and, yes, there will likely be an increase in property values as well as a greater demand for new land developments, new housing and recreational properties. Everyone on PEI, either directly or indirectly, will benefit from this project." The "everyone" mentioned in this brief is certainly an overstatement. The last time tourism was promoted in such terms was with the Development Plan of the early 1970s. Lorne Bonnell, the minister of tourism at the time, boasted that tourism would soon overtake farming and fishing to become the "number one" industry on PEI. At that time, the federal government was providing money to initiate the promotion, and all Islanders were to become "Sidewalk Ambassadors."

In addition, many tourists were enticed to buy a part of PEI. A report to the legislature in 1971 stated: "If acquisition of land by non-residents was to hold at the present rate of approximately 22,000 acres per 5 years, some 204,000 acres, or 14.5% of PEI would be foreign-owned by the year 2000." A large proportion of our coastline was lost to local ownership and control, and converted into summer homes for non-residents. Legislation was introduced in 1972 to limit the amount of acreage and shore frontage that non-residents could own without Cabinet approval. Cabinet has not been very active in enforcing these limits, however.

One of the big tourism plans of the 1970s called for the creation of two National Parks, one in Eastern Kings and one in West Prince. When local people discovered that the government was expropriating farms for the needs of non-residents, there was an outcry and both parks were halted. In 1977, an appeal to the Land Use Commission by the National Farmers Union regarding agricultural land expropriated for a Tourist Information Centre was successful. By this time, government was getting the message: Keep tourism in its proper place. It began to recognize that tourism initiatives must proceed carefully or the industry would lose the goodwill of the public upon which it depends. Islanders have been increasingly on guard against some of the "colonial" aspects of tourism, which threaten the culture and dignity of people who work in the traditional resource and service economy. Some capital-intensive projects and proposed land swaps, like the Greenwich Development, have been halted. Others have gone ahead because there has not been a clear willingness on the part of government to implement a policy. Great energy is needed to deal with mega-project proposals as they arise, one by one. The inability of successive governments on PEI to develop a working policy has limited the ability of Islanders to effectively plan for the types of resource development that we would like to see.

If free trade remains in place, it will become increasingly difficult to enact legislation and plan for the type of province we would like to live in — the enactment of so-called "private property rights" will severely limit us. Our present legislation has been challenged by non-residents in the courts. If we are shackled from having the right to protect our resources and culture, and if a private global concern controls the vital transportation link, we are working against heavy odds. The Landlords of the Strait will be in a powerful position to influence government and control the growth and direction of the tourism industry. Will they move it in the direction of the hard sell, Coney Island commercialism that seeks to attract millions of tourists? If the goal is for fast growth, this type of hyper-commercialism is the only option open to them.

The contradiction develops when many tourists like a place and want to buy land (at prices above what farmers are capable of paying) for retirement homes, weekend retreats, summer cottages or hobby farms. Ironically, the more people who do this, arriving with highly urban backgrounds and excess wealth, the more they end up destroying what attracted them in the first place — the diverse, resource-based, rural economies of farming and fishing.

We do have options, and alternatives to the proposed development scenarios outlined above. They are based on lived experiences, and informed by a strong sense of maintaining and using available resources. We are fortunate that steps have been taken towards conservation in the lobster fishery, with the implementation of trap and season limits. The lobster fishery is the cornerstone of the fishing industry, and the future effect of a bridge on the fishery must be better understood. For years, Island fishers warned of damage to the ground fishery inflicted by the dragging practises of industrial fishing. Now, the ground fishery is so depleted that it has almost closed down. The cultivated shell fishery, however, continues to show great potential, because of the water quality of our rivers. It requires continued protection from the effects of the erosion that is part of industrial agriculture.

Within agriculture, we must get away from the desire for an industrial fix, and create ways in which ecologists and farmers can work together. This must be a gradual process with the goal of redesigning what we do and how we do it. Our supply management system within the family-size operations of the dairy industry have made this type of farming the most stable force in rural PEI. Applying a similar marketing approach to vegetables and potatoes could help us to sustain and continue to develop high quality vegetables. A healthy, quality product is what consumers are demanding, and we have the soil and climate to meet this demand. We must build a sustainable foundation that works with nature to conserve and protect resources and minimize waste.

A more enlightened model of tourism could create partnerships that would conserve local environments, instill respect for cultural traditions, and build interactions between people. Such an approach would result in a sustained growth in the tourism industry, not the destructive, disruptive, rapid growth in volume that would eventually result from exchanging our land base for the capital benefit and profit of the transnationals who wish to build a fixed link.

There is a pattern to our history, and in our land. We can learn much from the experiences of previous generations, and, if we are alert and persistent in our efforts, work to build ourselves a future of which we can be proud.

From Public to Private: Privatization and the Link

Sharon Fraser

When you live in Prince Edward Island, your life rhythms are attuned to the ferry schedules. Arrival and departure times of the ferries — including hours of waiting time — become part of your consciousness, announced, as they are, on radio stations as regularly as weather forecasts and time checks are in other places. And when your ferry docks at Borden, you're home — even though you might have another hour- and-a-half drive to get to where you live.

The rhythms of Island life and the blessings of islandness are issues in any discussion of a bridge to the mainland. The fishing industry is an issue.

The environment is an issue. The loss of ferry jobs is an issue. But, so far, privatization of transportation across the Northumberland Strait hasn't been much of an issue. Why is that?

Consider this: When the bridge is built, the crossing between Cape Tormentine, N.B. and Borden, PEI, will be in private hands for the first time in history. In the simplest terms, the government is going to pay a great deal of our money — $42 million a year for 35 years — to a private company to take over an essential public service. In the future, the toll we pay will not go to support Island workers and to provide a safe and efficient passage across the Strait; instead, much of it will leave the country in the pockets of Morrison-Knudsen of Idaho, U.S. and GTM International of Santerre, France, two of the companies that make up the Strait Crossing Inc. (SCI) consortium.

Most Canadians have always believed that there are services which must be available, affordable and accessible to all people. These services are unlike goods and should be provided on a non-profit basis. Among those services, most Canadians would include education, health care, electricity and postal service. Most of us would also include roads, railways, ferries and — yes — bridges on our list.

The Royal Commission on National Passenger Transportation, which issued its interim report in 1991, says many Canadians believe that government should provide transportation infrastructure. "This would include ... building and maintaining highways, bridges, airports, railway tracks and harbours. Proponents of greater government ownership are concerned that private industries, which operate to make a profit, will not have the motivation to ensure that social, safety and environmental goals are met ..."

But something odd has happened. A very small minority of Canadians — aided and abetted by both federal and provincial governments and a vigorous business press — has decided that many of these services should be turned over to the private sector and used to make profits. And somehow, gradually, with a clever mix of misinformation and Orwellian newspeak, they have convinced many Canadians that there is something beneficial in taking public services out of the public's hands and putting them into the hands of private enterprise.

They have convinced Canadians that there is something deviant about public ownership of services and they characterize those who speak against this transfer of responsibility as being part of another age — wide-eyed innocents, unable to keep up with the demands of their global economy, doomed to be left behind in a vanished world where service and people were still more important than profits.

These perspectives of the powerful spread like wildfire during the '80s — the decade of greed — and, supported by the loud voices of government and their allies in business, they stopped being perspectives and began to be perceived as truth. In boardrooms, in cabinet rooms and in newsrooms, the idea of dismantling the services on which our country was founded became the norm. Other points of view which spoke of the need for public participation in the provision of essential services were either ignored or considered abnormal — perverse, even.

Privatization of public services grows out of ideology and thrives on mythology: the mythology of debt reduction, of efficiency and the importance of competitiveness, and of the triumph of economics over politics. Reducing the debt has become the ideal for the 1990s, as politicians and business people trip over each other in their zeal to cut government costs, reduce the size of government, and mostly, to sell off government services. They expect no argument: After all, how can anyone argue that debt is good?

The Minister of State for Privatization and Regulatory Affairs, John McDermid, plays down — although he does not dismiss — deficit reduction as the major reason for privatization in Canada, claiming instead that selling off government services reduces the number of demands on government and adds to the corporate tax base.

"Crown corporations are usually carried on the government's books at cost," he wrote in *Canadian Business Review* (Winter, 1989). "Therefore, the impact of a sale on the deficit depends on the difference between the sale price and the value recorded on the books. That number can be positive but often it will be neutral or even negative."

He goes on to say that privatization allows government to govern, by reducing the time and resources it spends on the management of Crown corporations. "Privatization also reduces the demands being made on the public treasury," McDermid wrote. "As governments have become more concerned with bringing their deficits under control, public-service managers have faced ever-increasing pressures to limit spending ... Selling Crown corporations is one way of reducing the demand on government for capital and resource requirements and freeing up scarce government resources for other purposes."

Herschel Hardin, in his book *The Privatization Putsch*, (published by The Institute for Research on Public Policy in 1989) takes issue with the premise that deficit reduction or saving government resources are relevant arguments. His research into the sale of public enterprises and services in Europe and in North America demonstrates that privatization has had little or no effect on the debts of the countries he studied:

> Trying to come to grips with the privatization arguments as they surface in Great Britain and are repeated by acolytes in Canada is like wrestling against a particularly slippery tag team. Once one

argument is defeated, it isn't important any longer; the reason being touted for privatization is an altogether different one. Pin that other argument down and yet another one is cited instead. Since the Thatcher government began its privatization drive in the early 1980s, the tagging of reasons has gone around full circle, and ones that bobbed up at the beginning and were discredited are now bobbing up again.

One of the early reasons for selling off the nationalized industries was to cut a dent in the annual government deficit ... The same reason was bandied about in Canada, too. Margaret Thatcher was going through a tight-money monetarist phase where the deficit was considered to be evil incarnate and had to be reduced. What better way than to sell off publicly-owned companies and use the cash?

The late Harold MacMillan, a true-blue Conservative disdainful of Mrs. Thatcher's rationalization, said that selling off the state assets would no more provide a lasting economic solution than selling off the family silver. This wasn't all that was wrong with the policy. The one-time effect on the deficit from the sale of assets was undermined by the loss forever of annual income from those assets. Even academic theorists of privatization dismissed the reduction-of-the-deficit ploy as illusory. There is no "free lunch" when it comes to reducing the deficit, said one. (George Yarrow, "Privatization in theory and practice," *Economic Policy*, April 1986, p. 360).

Efficiency and Competitiveness

"Selling the government's interest in corporate holdings improves the efficiency of the organization privatized by subjecting it more fully to marketplace forces," wrote John McDermid in 1989. "Freedom to access private-sector funding and the need to achieve adequate returns tend to make managers focus more closely on cost control, quality service to their customers and bottom-line results."

Of all the reasons given for privatization of Crown corporations and public enterprises, efficiency is cited most often. In what Herschel Hardin calls a "rigged debate," the concept that a private company is run more efficiently than a public enterprise or a Crown corporation is simply not questioned.

J.E. Konrad, writing in *Optimum* in 1989, says:

The argument for privatization is that greater efficiency can be achieved because the profit motive encourages better management of resources. Incentives are commonly used in the private sector to reduce costs. Furthermore, management is less prone to avoid confrontations over cost issues. As well, labour rates in the private sector are closer to those established by the market and show less rigidity on the down side. For example, governments tend to resist adding to the unemployed at a time when job creation is a priority. The private sector is quicker, therefore, to react to change and is free from many of the constraints imposed on public enterprises.

Leaving aside the fact that these are arguments in favour of creating a low-paid, insecure workforce, do they have any validity for the user public? In Margaret Thatcher's Britain, after the orgy of privatization in the '80s, a survey of companies after privatization disclosed that product quality and customer service ranked 10th and 11th out of 11 items. The same survey found no strong evidence that privatization had improved profitability. The effectiveness of competition can also be questioned.

Herschel Hardin, writing about Great Britain in the 1980s, says:

> A few sceptics began to look around them to see what did indeed happen when public enterprises were involved in competitive situations ... If labour costs or fuel costs were higher for one company than for another, and its profits were correspondingly lower, this did not mean it was less efficient or its employees less productive, only that it paid more for certain things.
>
> One of the studies that kept being mentioned, even in Britain, was on Canadian railways. It showed that publicly-owned CN in competition with privately-owned CP was every bit as productive — indeed, from a considerable handicap, it had slightly passed CP in productivity by the final years of the study.

John McDermid — on behalf of his government — advances the belief that decisions in privatized companies are made on the basis of economic and not political considerations. His view that bottom-line policies must be pursued at all costs fails to take into consideration the socio-economic effect that businesses — either public or private — can have. Marine Atlantic, for example, employs 651 people, much of whose income goes to support the PEI economy. But Marine Atlantic also supports other businesses in the area — by buying from them or, in the case of gift and hospitality shops, by delivering them customers.

In John McDermid's business world, this kind of economic support system is not a priority:

> Crown corporations were originally established to achieve specific public policy objectives, but they are today increasingly required to look to the bottom line. These two goals are sometimes incompatible. For example, the question arises: Should a Crown company be required to buy from a regional supplier in an effort to promote economic development, when the cost of those supplies is higher than it would be from a competitor outside the region? The conflict created by incompatible objectives can have a negative impact, both on the way in which public policy is carried out and on the company's ability to achieve its commercial objectives. In the absence of ownership, the government implements public policy through the use of other tools, such as regulation, tax policy and spending power.

But the triumph of economics over politics — even if that could be proved to be a good thing — is another myth. The people who control the money (and there are fewer and fewer of them) also have an unhealthy influence on the politics of our country. Herschel Hardin writes about the family dynasties — the Reichmanns, the Irvings, the Bronfmans, the Westons, the Thomsons and others — who hold a significant interest in more than 500 companies in the country:

> The large chartered banks have grown more powerful, with mushrooming financial assets, and are now taking over investment-dealer companies. There is growing cross-ownership between financial and non-financial companies (in the form of trust companies) on top of the old interlocking boards which tie the banks to their major conglomerate customers. The tax dodges which have helped them build up these empires at public expense have become notorious.

> With that economic power, too, indirectly comes political muscle ... the chairman of the Ontario Securities Commission, has argued that big corporate conglomerates "exercise power that extends far beyond their obvious function as efficient producers of goods and services ... the large corporation yields power that controls, directs and influences large segments of society." ... This ability carries with it leverage over the terms of investments: the level of pay and benefits; the kinds of jobs; how the work is organized; what subsidies will be required ... ; who gets supply and service contracts; even safety and environmental factors.

The arguments in favour of privatization don't bear careful scrutiny, particularly those myths that promote the twin concepts of "efficiency" and "competitiveness." These are code words for businesses that wish to increase revenues by cutting expenditures. Privatization has always been about profit — and that means providing services as cheaply as possible. Sooner or later, in the need to continue increasing profits, costs will have to be lowered further; lower costs mean lower wages for employees and poorer service. Sooner or later, cost-cutting in a business that puts profits before people leads to lax safety and environmental standards, unsafe working conditions and the loss of union representation for workers — as well as the loss of job security, adequate pensions, or any other benefits.

But if profit is the motivation for privatization for private sector companies — and it is — what are the motives of those politicians who have worked so diligently over the past decade to deprive the Canadian people of our publicly-owned services and infrastructure? Surely politicians have two main incentives to favour privatization.

The first is that it divests them of their responsibilities to the people who elected them. When public services are privatized, there is no political accountability at election time. If tolls are too high, if service has deteriorated, it's the fault of the private sector, so there's no need for taxpayers to consider complaining to their MLA or MP or voting him/her out of office. And, on the other hand, those politicians who favour privatization can cite the mythology which says that private business is more likely to be efficient and debt-free, more likely to add to the economy through job creation and payment of taxes, more likely to reduce public debt. The mythology is wrong: Privatization doesn't create jobs, it doesn't reduce the debt and it doesn't make any change in tax revenues.

Today, when we cross the Strait, our toll goes to the Crown corporation Marine Atlantic, where it is added to the federal subsidy to provide us with a safe, reliable, environmentally-sound, well-staffed service. A few years from now, when we cross by bridge, our toll will go into the coffers of a company whose first responsibility is to its shareholders, not its customers. That's the nature of business.

Meanwhile, the people of Canada have invested billions of dollars in their public corporation, have watched it function efficiently over the years, and must now watch as it's taken from them and turned over to a private enterprise — two-thirds of which is from outside the country. Moreover, we are being asked to pay more for a basic need, in order to increase that company's profits.

As public enterprises become profit-motivated private companies, the former owners — the taxpayers — have already paid a high price. When we collectively own our essential services, we are all participants in a true shareholders' democracy. We have a say — through our elected representatives — in how those services shall be run and presented. We have the potential to use those public enterprises as a centrepiece of a mixed economy

— a place of stable employment where workers at all levels are responsible to the community.

When we rush headlong into privatization in a country where the only definition of "business" has become "the art of making money," we lose that potential for a more secure economic future. We lose control over one more sector of our economy. We allow a small self-interested group to purloin what has always been ours — in this case, a service that was promised as a condition of Confederation — and then allow them to make us pay for the privilege of using it.

Thanks to privatization propaganda, we've accepted this as normal and have allowed it to be a non-issue in the debate surrounding the SCI bridge.

The Media and the Fixed Link

Martin Dorrell

The fixed link debate that has preoccupied Islanders for the best part of a decade began without fanfare as a small item tucked into the pages of the April 3, 1985 edition of the Charlottetown daily, the *Evening Patriot*: "OTTAWA — A group of private-sector entrepreneurs has made a proposal to the federal government to build a causeway between New Brunswick and Borden, PEI, says Hillsborough MP Tom McMillan."

In a CBC Television documentary three years later, McMillan said he'd come to favour the link after considerable thought, but any dark nights of the soul must have occurred before this somewhat contradictory observation: "Without committing myself to the principle of a fixed crossing, I think it could be sold as one of the modern wonders of the world. It could enhance the Island's attractiveness as an island and could be a tourist attraction."

Although court battles and the search for financial backers lay ahead, the story ended — for the purposes of this discussion — almost eight years later, on Dec. 3, 1992: "OTTAWA (CP) — The federal government will press ahead with plans for a bridge to Prince Edward Island, Finance Minister Don Mazankowski said Wednesday.

"Advanced engineering and environmental work on the fixed crossing 'will be undertaken to allow construction of this $800-million project to begin in the spring,' Mazankowski told the Commons in his statement on the economy."

The tale that unfolds between these two stories involves dozens of journalists, whose clippings fill 12 bound volumes and whose reports amount to more than 500 CBC Television videotapes, from 90-second clips to half-hour documentaries.

It's a story combining immense technical complexity with political intrigue, economic arguments with environmental subtleties. And yet, at the end of the day, it's profoundly emotional. Before and after all the rhetoric, all the studies, all the debate and all the coverage, most people responded to the proposition of a fixed link from the gut. For some, it's progress, an inevitability, the Island's only hope — for others, an environmental, economic and social catastrophe.

But did all the ink and videotape clarify issues — or muddy them? Did the media simply react to noisemakers or did they also strive to add meaning, detail and context through their own investigations? Did the media contribute to the debate or merely convey it?

The media are production-oriented, deadline-driven, short-staffed and turnover-plagued. They prefer lively quotes and colourful personalities to technical detail and abstract thought. That — combined with the stark fact that there was no honest broker, no objective source of information — made this the challenge of a lifetime for an Island journalist. The clippings and tapes document their response to that test.

• • •

A flurry of stories follow McMillan's remarks, as supporters surface and editorials express skeptical interest. "Will sheer economic desperation one day buckle the knees of the most romantic Prince Edward Islander," the *Patriot* muses, "and give birth to the long talked-about, fought-about causeway?" And, the next day: "The province's very inability to negotiate reasonable ferry service adds fuel to the causeway argument." The Summerside daily, the *Journal-Pioneer* won't rule out such a project, "but if it happens, it will be in the lifetime of only our younger citizens."

In September 1985, the *Guardian*, the other Charlottetown daily, comes out in support of the link, citing convenience and the prospect of jobs. "But would the province be any less an island because a tunnel connected it with the mainland? Is it any less an island because of the electric power cable under Northumberland Strait? Hardly."

The debate remains low-key and spasmodic until the fall of 1986, when the first vague proposals are unveiled. Intrigued, the *Patriot* says a link would boost the Island's prestige and competitiveness, but it wonders about public support because "a mile of pavement is a big project for Islanders."

At this time, a *Guardian* reporter interviews an official with the Mackinac Bridge Authority in St. Ignace, Michigan, about the impact of the 8 kilometre bridge on tourism and the economy. He learns that traffic increased more than 75% in the first year that a bridge replaced the ferry. This story, modest as it is, remains one of the few in the Island dailies to offer more than routine coverage of events.

The *Eastern Graphic* in Montague is pressing McMillan to call for a plebiscite. The weekly says the link is attractive to Ottawa because it won't have to pay for it, and can eventually drop its subsidies. However the newspaper argues it would cost ferry jobs, put pressure on land and tourist facilities, boost mainland businesses and pose transportation problems in winter. It wants to know where the provincial government stands in the debate.

The *Guardian* concedes that, if improved transportation results, it may be a mixed blessing. "Who were the winners and the losers when the automobile and paved roads made for easy transportation between Charlottetown and rural PEI? Just as rural communities struggled to retain their unique character as business and populations gravitated to Charlottetown and Summerside, so does the Island community risk losing its identity to the bigger mainland shopping centres." By the end of the year, the *Guardian* agrees there should be a plebiscite and, if there's support for the idea, more studies.

Meanwhile, CBC Television is outlining the various proposals, including the unforgettable comment of one would-be builder that a link might help the fishery since fishing is always good off a bridge. In December, the CBC attracts large crowds to its public forum with three potential builders, and broadcasts the results in two half-hour segments. Transportation minister Robert Morrissey warns that the outcome of any plebiscite must be unequivocal to warrant action. "Certainly we would be looking for a very clear, strong consensus from the Island residents — certainly 75-80% of Islanders supporting the fixed crossing." Those remarks should come back to haunt Morrissey. They don't.

By year's end, Public Works Canada (PWC) plans several feasibility studies. Late in the fall of 1987, information from those studies is released, a dozen information meetings are held and Premier Joe Ghiz calls a plebiscite for January. The studies are outlined in newspaper reports, sometimes in mind-numbing detail, but they're hard to digest. The *Guardian* says the information is too massive to prompt an intelligent vote, although a few days later it concludes that one can't ever get complete answers to good questions. Editorially, it refrains from asking any.

PWC releases two studies only after the *Graphic* publishes extracts that the department says are distortions. By now, if there was ever any doubt, it's clear PWC is actively promoting the project. The *Graphic* urges Islanders to vote No, arguing that McMillan should be appointing an Environmental Assessment Panel (EAP) to examine the idea.

CBC's coverage for much of the year is unremarkable except for two lively studio debates late in December. PWC spokesperson Glenn Duncan is unequivocal about the cost of a fixed link. "All the builders have told us is that they can build a fixed crossing for less than the cost of the ferry service," Duncan says. "If that is not the case, it will not be built."

As Islanders prepare for the crucial vote, they do so with little easily accessible information. Media reports of the studies are sketchy or confusing. No one attempts to gather experts to prepare informal impact studies or to comment on the validity of the official ones.

Both the *Guardian* and the *Patriot* urge a Yes vote to, as the *Patriot* puts it, "keep the dream alive." The *Journal-Pioneer*, usually loath to express an opinion, says a No vote is fuelled by fear of the unknown. But *Graphic* publisher Jim MacNeill says the media bears part of the blame if much is unknown. "None of the media did a completely effective job ... Studying those reports took a lot of time and effort. Even further effort was needed to put the information into an understandable form. Yet it should have been done. Polls and phone-ins aren't a substitute for plain simple facts by reporters. The news media, or at least the press, should also have been pointing out the inadequacies of the consultants' reports."

In the days before the vote, the *Guardian* contents itself with surveys of the opinions of businesses and commodity groups. CBC-TV does a more coherent, effective and user-friendly job in a series dealing with the possible impact of a link on the fishery, farming and the economy in general. Fishers are pressing for a tunnel, which poses less threat to their livelihoods. PWC's Duncan, who has said earlier that a link wouldn't be built if fishers are opposed, is now hedging ever so slightly. "I would think," he says, "it would be extremely difficult if not impossible for a project to proceed" without the fishers' support. However, he is unaware of a single bridge that has caused environmental damage. Bridges, he observes, "are just benign."

After Islanders narrowly vote Yes, Ghiz tells the CBC that work will proceed to establish a fixed link unless something unforeseen occurs. No one recalls Morrissey's insistence that any vote would have to be overwhelmingly positive.

The *Journal-Pioneer* quotes McMillan as saying that an environmental review would "be a very lengthy process. Islanders must ask themselves if it is necessary because it could very well set the process back at least two years." Only the *Graphic* finds this a curious stance, coming as it does from a federal environment minister. The weekly continues to push for a review. McMillan asks to be relieved of any environmental responsibilities associated with a link.

In March, the *Graphic* discloses that PWC is sending a delegation to look at bridges in Scandinavia. An ATV cameraman and reporter go along. ATV pays for the reporter, but not the cameraman who — in a decidedly odd deal — shoots footage for PWC in exchange for a free ride. One of the bridges the team investigates links the small island of Öland to the Swedish mainland. The comparison is a little strained. The bridge crosses water half the width of the Northumberland Strait, currents are weaker and milder winters produce little ice.

The CBC pays its own way to examine the bridge — perhaps its most ambitious undertaking throughout the period. The result: a major documentary and some spin-off pieces. As I remember it, the documentary was extremely enthusiastic about the bridge. Perhaps that memory was coloured by some off-air comments I heard in the halls of the CBC at the time, but, today, I find the documentary quite fair and balanced. On the plus side, islanders said the bridge was more convenient, it had created new businesses, made farmers more competitive, had not harmed the fishery and "there's not an Ölander to be found who would give up their fixed link."

On the other hand, the documentary cited a ten-fold increase in visitors (and not always welcome ones), prompting belated land-use policies to control development after cottages outnumbered permanent homes. Heavy traffic created chain collisions and occasional queues of several hours. Few new industries or year-round jobs were created. Small retailers were hurt by larger mainland competitors, some services were centralized off-island and many residents no longer considered themselves islanders. Indeed, it's hard to fathom why an Ölander couldn't be found who would give up the bridge.

By the fall of 1988, seven contenders for the project have been winnowed to three. With an election in the air, the enthusiasm of the public works minister for the project borders on the rabid. The *Guardian* quotes Stewart McInnes vowing to an Island audience "there will be no unemployment on PEI for anyone who wants a job on the fixed link project." He promises Islanders 7,000 direct jobs and 13,000 spin-off jobs. And, he says, construction could start in the spring of '89. Three weeks later, the newspaper cites a report commissioned by McInnes's department predicting 2,500 direct jobs arising from construction — only half of them for Islanders. No one points out the contradiction.

Fishers are furious that a tunnel hasn't made the short list. The *Patriot* is unsympathetic. It says Ottawa has spent $5 million on consultants who say a bridge poses no risk to the fishery, yet fishers don't trust these experts. "Are Island fishermen refusing the long drudgery of study and deep thought?" the *Patriot* asks. One might, at this point, ask the same question of the dailies — certainly, of their editorials.

"If the dogmatists had their way in this province," the editorial thunders, "we wouldn't be part of Canada, have school consolidation, a provincial university, a provincial hospital, mixed marriages ... or fishing boats with motors."

A recurring theme in the editorial pages of the *Patriot* and the *Guardian* portrays the federal government and the fixed link as benign and progressive, and the link as an inevitability, the key to long-term economic growth. Its opponents are Luddites who seize any issue, no matter how flimsy, to stop it. Proponents, presumably, refrain from seizing any issue, no matter how flimsy, to promote it. Opponents are cringing whiners, romantics fearful of change, intellectuals out to protect their own narrow interests. Proponents are fearless supporters of change, committed to ending the Island's dependence on transfer payments. The burden of proof tends to rest with opponents. As Jack McAndrew writes in the *Graphic* at this time, fishers must convince the government that there will be environmental damage to the fishery — but the federal government need not convince fishers a bridge won't destroy their livelihoods.

Despite their support for a link, all three dailies express occasional reservations about the process. The *Guardian* says the province has "a moral obligation" to hold a second ballot since voters thought Yes would lead to more information, not a bridge. The Journal-Pioneer wants to know what happened to the tunnel. "The proponents of the fixed crossing should be concerned to minimize the opposition," it says. "They are not doing so by refusing to say why the tunnel proposal was rejected."

Only three weeks after denouncing fishers for their lack of faith in experts, the Patriot does a complete about-face. It puzzles over a PWC study that says spring ice break-up may delay but not hurt the fishery, yet concludes that the environmental factors are so complex that indirect effects can't be predicted. "What kind of self-respecting business operators could be expected to accept this kind of double-talk and imprecise analysis?" the *Patriot* roars. "Fishermen are justified in their fears, and in insisting on more convincing answers."

The *Graphic*, meanwhile, receives honourable mention at the Michener Awards for meritorious public service in covering the fixed link issue. As the newspaper continues to press for a full-scale Environmental Assessment and Review Process (EARP), McMillan responds that the proper process is being observed — a full environmental review follows the selection of a specific project.

As the year ends, both McMillan and McInnes are sidelined, losing their seats in the general election. In January 1989, Ottawa ends any talk of a fast track by abruptly announcing a full-scale environmental review. The CBC reports, unequivocally, that if the Environmental Assessment Panel (EAP) finds a bridge would harm the environment, "the whole project will be cancelled." The report says that although the federal government is declining to explain its sudden change of heart, it is apparently moving to satisfy public concerns. As a result, a specific project may be more than a year away. The dailies reluctantly conclude that a review is necessary to clear the air. McMillan suggests in a CBC interview that McInnes's prediction that construction could begin in early '89 galvanized opponents, forcing the government to hold an EARP before a specific project was selected. A fast-track supporter himself, after attributing the delay to a former colleague, McMillan suggests biased media coverage is partly to blame. If he elaborates, we don't hear about it.

The selection and background of the panelists receive little or no scrutiny. Neither does the timing of the appointment of the review Panel — too late to look at alternatives such as a tunnel or improved ferry service, yet too early to examine a specific bridge proposal. The *Journal-Pioneer* asks whether it isn't "somewhat strange that the government would have only discerned [some Islanders' concerns] now?" It doesn't pursue the question. Nor do the media provide much history about 33 previous environmental review panels, their makeup and decisions, or the response of the government to their recommendations.

Hearings in June shed light on a couple of issues. Federal fisheries scientists say PWC's ice studies are inadequate. And PWC says a tunnel was too risky, construction didn't offer the region enough short-term jobs and the bidder wouldn't drill a $50-million exploratory tunnel.

Two months later, the review Panel calls for several additional studies, including ones to address ice and the impact of increased traffic and development on PEI. While the studies proceed, media interest appears to wane. They aren't attempting to conduct their own investigations. Then, as 1990 begins, the *Patriot* is having second thoughts about the EARP. "The province," the newspaper grumbles, "seems content to let a review board and others make the fixed link decision."

Even at this late date, something as elementary as the estimated cost of the project is a matter of confusion. In the space of a week, Canadian Press refers to it as a $600-million project, then as a $750-million development. The media cover the new round of late-winter public hearings in exhaustive detail, events that produce bountiful copy with little investment. At one hearing, Marine Atlantic argues that it operates the ferries for smaller subsidies than Ottawa is allocating for the link. Public Works Canada has said repeatedly that a link won't be built if it costs taxpayers more than the ferries. If this information turned up in basic research by the media beforehand, I missed it. It's a point that gives rise to new anti-link groups and to arguments that dog the issue for the duration of the debate. Now it provokes little editorial comment or investigation.

In August, the EAP Report rejects the link. It lists several reasons: A bridge is irreversible, it isn't sustainable development; the adequacy of any compensation plan for more than 600 displaced ferry workers is doubtful; there is no industry to replace the almost $25 million that the ferry service pumps into the Island economy every year; and there are serious questions about the impact on the Island's tourism industry and on its roads.

The EAP Report also rejects the project because it's not satisfied that ice wouldn't be caught in the piers of a bridge during spring break-up, remaining in the Strait for as much as two weeks longer. It could scour spawning grounds, threatening the $75 million lobster, scallop and herring fishery. It could also delay the spring lobster fishery, dramatically reducing the rate at which lobster mature. The panel suggests no compensation program could ever cover the long-term losses to the fishery. It also says that a delay of only two days in the ice leaving the Strait would be acceptable. The Panel itself says that it should have been appointed sooner — or later.

McMillan isn't happy, questioning the makeup of the Panel in a *Guardian* interview. Who are the panelists? How did they get there? Interesting questions, interesting timing — coming as they do from a former environment minister.

The *Journal-Pioneer* pronounces the fixed crossing dead for another quarter of a century. The *Guardian* urges Islanders to push for better ferry service since Ottawa "would be foolhardy to ignore its own panel's opinion on something as sensitive as the environment." But the *Graphic* says the link is far from dead. And soon it's clear that PWC isn't about to bury it.

After a two-month hiatus, PWC minister Elmer MacKay appoints a new panel of ice experts to re-examine earlier findings. The *Patriot*, despising intellectuals while somehow maintaining deep faith in experts, is delighted. "The ice panel will come up with a solution," it assures readers. "Experts couldn't call themselves experts if they can't solve problems."

Former EAP panelist Carol Livingstone tells the *Journal-Pioneer* that there are concerns other than ice. And she despairs of the process. "It looks like they're going to keep appointing panels until they get the answer they're looking for." The question of "ice-out" appears the most tangible, the most specific, the one Public Works Canada can most effectively address. It concentrates on that and, as a result, so do the media. The media focus on action, events. When PWC acts only on the question of "ice-out," declaring it the only outstanding issue, that's what it becomes.

In the spring of '91, the new ice panel gives the project the green light. Continuing arguments from the critics exasperate the *Journal-Pioneer*. "No amount of evidence showing a fixed link has not destroyed the way of life for others who have had their island connected after years of separation has been able to convince these people. They are entitled to have their views and to air them, but do they really listen to the other side, or are they just unmovable in their theories because they don't want to be shown they are wrong?" The *Journal-Pioneer* does not cite the evidence to which it refers and, if the proponents of the link have proof that the project won't harm the environment or the province's social fabric, the newspaper hasn't presented it.

In April, the *Graphic* picks up on the disparity between the $25 million subsidy Marine Atlantic is receiving and the $40 million earmarked for the operator of a link. "The unfortunate thing," the *Graphic* says, "is that the Charlottetown news media have done so little research into the facts and fictions about the fixed link that nobody thought of questioning Elmer MacKay." Over the summer, the *Graphic* runs a series exploring the review Panel's unanswered questions, including its request that the province determine the optimum number of Island-bound tourists.

As 1992 begins, three bids are undergoing environmental scrutiny. Meanwhile, a *Graphic* story dealing with the impact of winds on a link creates a media gale as reporters trot out their own meteorological experts. What gets lost in the storm is the simple truth that no one knows the velocity of winds at bridge height over the Strait because it has never been measured.

As January ends, the three bidders pass environmental tests and are preparing their final financial packages. A CBC documentary updates viewers on progress towards meeting the 10 conditions the premier has laid down before he'll sign a deal. The media pay belated attention to the subsidy issue when a new group, Islanders for a Free Ferry Service, makes it an issue.

As Islanders wait for a decision and two bidders are dropped because their subsidy requests are too high, Canadian Press (CP) runs a backgrounder out of Ottawa that aptly illustrates the national media's infrequent and indifferent coverage. "Years before Prince Edward Islanders embraced the charms of Anne of Green Gables, they began a long-running dialogue over the merits of a fixed crossing to the mainland," CP explains. "The mammoth structure would replace federally subsidized ferry service between the points, reducing total crossing time to about 15 minutes from the current hour and 40 minutes."

And, in a reference that must particularly delight PWC, CP writes off any lingering opposition as clearly coming from the lunatic fringe. "Critics say a fixed link could spoil the intangible mystique of the Island and tamper with the psyche of its 130,000 inhabitants."

CBC-TV profiles the Calgary-based president of the front-running company, Paul Giannelia of Strait Crossing Inc. (SCI). He's charming. The next night, a documentary gives equal time to the efforts of link opponents, especially the likely environmental challenges expected to reach the courts in 1993. When SCI's bid fails to meet Ottawa's financial requirements, the two sides negotiate a tentative deal.

• • •

Over the years, the Island media has produced no shortage of material. Almost every news conference was covered, every press release received notice, every hearing was attended. But the media, especially the dailies, conducted little research and were too often content to report the findings of others—and, for the most part, that research came from a lobbyist. Public Works Canada promoted the project and paid for the studies.

Was Tom McMillan in a position of conflict long before he resigned as environment minister? What lay behind the timing of the plebiscite? What did a Yes or a No vote really mean? What happened to the tunnel? What did those PWC-commissioned studies say? How would other experts assess the quality of those reports? How many short-term jobs would a link create? Why was an environmental review panel appointed when it was? What's the track record of those panels? How are members selected? What would the link's impact be on tourism? How do the bridge subsidies compare with subsidies for a continuing ferry service? How have governments responded to concerns other than ice?

If report cards were handed out for media coverage, the grades would be uninspiring:

The *Guardian*, the *Evening Patriot*, the *Journal-Pioneer*: F. Too many reporters covered too many events at the expense of analysing findings or conducting their own research. The newspapers offered almost no background or analysis, accepting what they were given on face value. Their editorials tended to be smug when they weren't sophomoric. They did little to help their readers make informed decisions.

CBC Television: D. In addition to succinct and usually coherent coverage of events, CBC encouraged debate and offered some background and context. But over the years, more than 25 reporters helped tell the story, leaving questions unanswered, inconsistencies unexplored. What might the corporation have accomplished had the link been a full-time beat for just one reporter?

The *Eastern Graphic*: C. The *Graphic* confined coverage of events to its own backyard. Its ferocious editorial skepticism raised questions and spotted inconsistencies. In its frustration with the process, however, it occasionally ran stories from lobbyists that would have been more appropriate as letters to the editor. In the end, it was frequently perceived as the haven for every anti-link argument.

The national media: F. Where were they? When they did show up, it was to perpetuate a stereotype: the Island as Canada's cartoon province—a backwater with more than its share of eccentrics harbouring deep suspicions about progress. If there were serious environmental, economic and social implications, the national media rarely noticed. They missed the story.

Imagine how the tale might have been told had it dealt with a heavily populated island sitting in Lake Ontario near Toronto.

The Government, the Process and the People

"It seems to us that there is something inherently wrong *to have a mega-project, destined to have profound effects on the province as a whole, and on the Borden area in particular, planned almost exclusively by the Government of Canada, the potential developers and a collection of consultants … 'Monitoring' the activities of the Government of Canada, the developers and the consultants will not suffice."*

Douglas Boylan,
Everything Before Us: Report of the Royal Commission on the Land, October 1990.

It is humorous, in an ironic way — former Premier Joseph Ghiz demanding integrity of a process, while simultaneously suggesting to Public Works Canada (PWC) that it circumvent the federal Environmental Assessment and Review Process (EARP) recommendation that the bridge not be built.

Joe Ghiz survived his terms in office by playing the role of honest broker, refusing to say if he favoured the bridge or if he opposed it. Even when the decision to support the project was made by Executive Council, effectively silencing government members and avoiding debate of the question in the legislature, people believed that, above all, Premier Ghiz had integrity. Prince Edward Islanders were aware of a curious lop-sided stand by PWC, who even admitted in 1989 that they were "proponents" of the project. But Joe Ghiz, until his February 1993 confession-by-affidavit, fooled Islanders, and he's made them feel that they've been had.

The pro-bridge bias that has been so evident in this process could have been better challenged if verifiable comparisons amongst all the crossing options had been made, and made clearly. An "easily comprehensible diagram" is what the Environmental Assessment Panel (EAP) requested. PWC said it couldn't be done. In this chapter, we have included a proposal of how such a comparison could be done — and the results of our study indicate that, of the four crossing options, the bridge is by far the worst. Worse too, was the timing of the environmental assessment, coming, as it did, at the mid-point in the process. A Panel review earlier in the process would have determined the type of crossing option most appropriate for Prince Edward Island. As it was, the EAP examined a generic bridge design, and not the specific Strait Crossing Inc. (SCI) design.

"Friends of the Island," a coalition of environmental, labour, fishers' and social action organizations and individuals, challenged a grouping of three federal departments, the governments of New Brunswick and Prince Edward Island, and the developer on this issue — and won. The federal government was ordered to hold an environmental assessment on SCI's specific bridge design and PWC was reprimanded for witholding crucial information from the 1989-90 EAP.

Now, if Islanders could have a choice of crossing options, too, much of the damage and divisiveness wrecked by this process could be undone, and maybe even forgotten — except in the history books.

Environmental Assessment: What It Is, What It Isn't and What It Could Be

Daniel Schulman

The proposed bridge between Prince Edward Island and the mainland is a mega-project. As things presently stand, environmental assessments of mega-projects are severely inadequate. But they don't have to be — the dream of a meaningful assessment process keeps environmentalists active in the struggle to change things. In this article, I would like to discuss three aspects of the process of assessing the fixed link: what has happened, difficulties in assessing mega-projects and what should really have happened.

What Has Happened

Let's start with a simple and obvious point. There are five ways to provide transportation across the Northumberland Strait: the existing ferry service; improved ferry service; a road tunnel; a rail tunnel; and a bridge.

Given the opportunity to change the mode of transportation across the Strait, how would a reasonable person decide what to do? That person would take the time to consider all of the five options with equal intensity. He or she would weigh the pros and cons of each, and then make a balanced and informed choice. This sounds so obvious, it is hard to imagine any other way of doing things. Unfortunately, this rational and obvious method is only a dream. It is most definitely *not* how Public Works Canada (PWC) has handled the Northumberland Strait crossing question. Let's briefly look at some key aspects of that process.

In the history of a mega-project such as the fixed link, there are three distinct points in time when a full environmental assessment could be called. Lets call these three points, A, B and C. Point A would be very early on, when the big questions can be asked — questions such as "why are we doing this?"; "what are all of the alternatives for achieving this goal?"; and "what is the best alternative?" In the case of the fixed link, this stage would have entertained all five options for transportation between PEI and the mainland with equal objectivity.

Point B would occur well enough into the process that one alternative had been generically selected, but no specific details had been established. Assessment at this stage could ensure that the early stages of the project design minimize negative environmental impact and/or maximize environmental enhancement.

Point C would occur when a specific project design is on the table. At this point, the technical details of the project, including specific design parameters and environmental monitoring plans, could be scrutinized.

In short, assessment at point A allows for choice and vision. Assessment at point B helps in anticipating major problems with a particular choice. Assessment at point C is a way to make sure the engineers have done their homework.

If a project is not subjected to assessment at stage A, we as a society have everything to lose. We lose perspective. We lose vision. We lose opportunities to make sound choices.

We talk a lot about "sustainable development" in this country. But only when we are mature enough to subject both our major projects and our policies to early stage environmental assessment (point A) will we be demonstrating a commitment to sustainable development. It is important to realize that not everyone loses from bypassing assessment at point A. There are always special interests who gain from this approach. In fact, it is usually vested interests that push our society into considering only one option (in this case, a bridge; in other cases, things like nuclear power or clearcut/replant forestry practices). When an environmental assessment is tacked on to the process after point A is bypassed, the process becomes just a tool for making environmentally destructive projects a little more bearable, not a tool for choosing the best option.

Of the three possible times to enlist the environmental assessment process, the middle option (point B) — the one that is too late to consider all alternatives but too soon to scrutinize specific technical details — is the worst. And it was this time that was chosen for the assessment of the proposed bridge. This timing increases the potential for vagueness and manipulation and decreases accountability. And, indeed, in the case of the bridge across the Northumberland Strait, the assessment has done exactly that.

It has ensured that, in the early stages, there was no open and independent attempt to answer the question "What is the best option for the people and ecosystem of the Northumberland Strait?" Instead, the outdated forces of political and corporate influence secretly answered that question for us. Only after this point, was the process begrudgingly opened up to a credible assessment away from corporate and political influence. This occurred in April of 1989 when the Environmental Assessment Panel (EAP) was appointed. But by then the choices were limited.

And even then, after this formal, credible assessment was undertaken and rejected the idea of a bridge (in August of 1990), the process was quickly wrestled back to the privacy of backroom influence. The fate of the input of fishery experts illustrates this point well. Of the two groups with the most credible opinions and studies on the possible impact of the bridge on the fishery — fishers and Department of Fisheries and Ocean (DFO) scientists — neither has been able to fairly and openly participate in any part of the process since August of 1990. This is despite the fact that both groups expressed grave concern about the project to the EAP, and provided input which significantly contributed to their rejection of the project.

Project proponents will debate this. They will tell you that DFO has been an active participant in the process all along. This is true. But as the PEI Fishermen's Association (PEIFA) so prophetically stated in their 1990 submission to the EAP, "DFO might in the future be subject to pressure from its sister department PWC to soften its official scientific position" (see PEIFA, 1990). Indeed, all DFO involvement in the process since 1990 has been moved to senior bureaucratic levels of the department and away from ground level scientists. Similar manipulation of involvement by fishers has occurred.

A properly run environmental assessment process is carefully designed to be outside of political or corporate influence. The EAP, which concluded its assessment of a generic bridge in August of 1990, was just that kind of process. Canadians currently spend millions of dollars each year on maintaining a credible process through the Federal Environmental Assessment Review Office (FEARO), but the public faith in that process is teetering.

Since August, 1990, there has been much fixed-link-related bureaucratic activity. It has all been undertaken in the name of environmental process: the "ice panel" appointed by Public Works Canada, the "environmental committee" comprising representatives of various governments, the "public information sessions" (often mislabelled "public hearings" and "public reviews," when they were neither) offered by Strait Crossing Inc. (SCI). Many people have been deceived into thinking all of these stages were part of an "environmental assessment process."

The fact is, no part of the process since August 1990 can qualify as environmental assessment. For those of us trying to engage in constructive debate, this has been an insurmountable point of confusion with many members of the public. Both the media and politicians have fuelled this confusion. All of this activity since August 1990 has been entirely directed by PWC, by senior bureaucratic levels in the various governments, and by the corporate interests involved. All questions posed and answers obtained since August 1990 have been directly subject to corporate and political influence.

So it is essential to understand that over the many years of debate and study, there has only been a small window of time during which meaningful, interactive public input in a non-threatening atmosphere was possible. This occurred between April of 1989 and August of 1990 — slightly more than one year of time. And the nature of the input was confined principally to discussion of the idea of a bridge.

This style of decision-making is very outdated. In fact, Atlantic Canada is currently living through the painful consequences of this old-style decision-making — the collapse of the East Coast fishery (the most recent wave of environmental devastation in Canada's Atlantic region) came out of a decision-making process that was directed by senior political and corporate influence.

Environmental Impact Assessment of Mega-projects.

Both the official FEARO environmental assessment (EARP) and the non-official PWC-directed evaluation of the proposed bridge have been based on very poor data. This is particularly true where evaluation has concerned the biological data from the Northumberland Strait ecosystem. The baseline data are, in fact, so inadequate that marine scientists believe we will never be able to attribute any long-term impacts to the bridge. I refer the reader to Rice et al (1989), CAFSAC (1989) and PEIFA (1990) for discussion of these matters. Even if we did have good baseline data, however, technical environmental impact assessment of mega-projects is, by nature, a very limited process. There are three very important limitations to the technical aspect of environmental assessment. I will call these three limitations "predictive uncertainties," "cumulative impacts" and "technically camouflaged values."

A mega-project is a special creature. An overwhelming number of environmental variables are involved. Each of them must be projected into the future. Some projections are made reasonably well and some are made very poorly. But with each projection comes uncertainty. And there are two kinds of uncertainty: known and unknown. Known uncertainties are the things computer models play around with. Unknown uncertainties are the things nobody plays around with.

Assessment of mega-projects is so uniquely worrisome because no one knows what happens when all of these uncertainties are combined. Do they multiply? Do they interact to produce new effects we never considered? These are difficult things to grasp. Yet many predictions resulting from the assessment of mega-projects are linked to other predictions, in a long sequence.

The fishers of Prince Edward Island illustrated this problem in their brief to the EAP with one tiny example drawn from a vast myriad of possible examples. According to PEIFA (1990), "models of ice dynamics require knowledge of ice hardness, which can be calculated from ice age, chemistry, and the manner of formation. However, ice hardness may change by up to 50% depending on the content of ice-inhabiting algae.

In the Northumberland Strait, changes in ice dynamics induced by a bridge may change conditions necessary for the growth of ice algae, which may in turn change ice hardness, which will in turn affect predictions of effects of the bridge on ice."

In addition, many impacts are cumulative. This point is very different from the uncertainties discussed above. Cumulative impacts are very difficult to predict. In the case of the fixed link, the original EAP noted that "considerable concern over cumulative effects remains." Let's just look at one example of cumulative impacts which would never have been predicted.

The first part of the James Bay Project involved the flooding of an expansive area of land in northern Quebec. Decay of the large volumes of suddenly submerged vegetation led to increased methylmercury uptake by fish. The James Bay Cree depended on fish. Now the Cree must eat grocery food instead. A combination of unfamiliarity with the European diet, the cost of healthy grocery food in Northern Quebec and the disorientation of having their world turned upside down has resulted in a Cree diet based on sugar and refined carbohydrates. Now, the James Bay Cree are developing obesity and diabetes problems on a social scale. Who would have ever predicted that obesity and diabetes in the Cree population would result from flooding? One shudders to think of the larger scale compounded or cumulative impacts of the James Bay Project which we have yet to understand!

Even with the best models and predictions of the day, and even if we did not have to worry about predictive uncertainties and cumulative impacts, there is another important limitation to technical environmental impact assessment of mega-projects — in the end, we entrust technical experts to make value judgements which have no special place with experts. Very few people realize this.

At some point in the evaluation of the proposed bridge, someone had to provide weightings to the various concerns. Someone had to give birds, groundwater, hotel operators, fish, realtors, fishers, potato processors, farmers, soil, plants, truckers, ferry workers, aquatic ecosystems and terrestrial ecosystems different weights. Someone had to decide that one concern was more important than the other. These decisions have been made under what I would call "technical camouflage." In other words, they have been made by technical experts who are not necessarily any more gifted at value judgements than anyone else.

People must understand that technical knowledge has its limits. Few scientists and technical experts are willing to educate the public on these matters. Many scientists and technical experts are, themselves, uneducated on the limits and social context of their work. And bureaucrats are, for the most part, entirely ignorant on these matters. Unfortunately, it is usually senior bureaucrats who administer decision-making processes which depend on technical understanding.

The eminent Canadian scientist and engineer Ursula Franklin offers appropriate comment in her book *The Real World of Technology*. "Today, scientific constructs have become *the* model of describing life around us. As a consequence there has been a very marked decrease in the reliance of people on their own experience ..." Repeatedly, over the past five years, local fishers, ship captains and residents have recounted natural events which the technical modellers have claimed are impossible.

What Should Really Have Happened

In my view, the evaluation of the fixed link has been nothing less than an environmental assessment fiasco. Taking this position, I must now offer a vision of what should have happened. There are two levels to this vision: At one level, I can offer a model of what could and should have been achievable within our present way of doing things; at a different level, I can offer a proposal for a future decision-making process that accommodates both technical expertise and dignified citizens.

What should have happened is discussed at the outset of this article: The five possible options for transportation across the Northumberland Strait should have been explored with equal open-mindedness. This could have been undertaken as a largely technical exercise. Many people maintain that such a process would have concluded that a bridge is the worst overall option, and that improved ferry service is the best option. We will never know the complete answer, because no one has ever been able to investigate the question.

At the second level, I suggest a process which is, perhaps, a little too advanced to have realistically been adopted here and now. But I offer it as a goal towards which we must move. This process would recognize the double-edged sword of technical expertise — both its importance and its limitations. It would also recognize the wisdom of experience-based local knowledge.

This process would involve early stage environmental assessment. It would clearly identify all alternatives for achieving the stated objective. Early public consultation would take place in an atmosphere untainted with any looming prospect of the tempting financial gains and contracts associated with one particular choice. There would be a clearly established and important role for technical expertise. But the participating public would also be openly informed of what technical expertise can and cannot offer. Indigenous knowledge and wisdom would be placed on an equal footing with technical knowledge.

Public consultation would also involve a creative facilitation process in a non-threatening atmosphere. Public consultation is poorly understood at present and rarely practised. It is all too often offered at a point very late in the decision-making process, and amounts to little more than public information on what has, for the most part, already been decided upon.

Some will argue that the general public are not capable of sophisticated analysis. I do not believe this to be true. What could be more technically complex than the issues surrounding the recent PVY-n potato crisis? Yet the PEI Department of Agriculture, Island media and the potato industry have recently managed to elevate understanding of this matter within the general public to a high level. Responding to similar arguments that the general public are not capable of complex decision-making, author and political scientist Noam Chomsky recently noted in *Harper's Magazine* how sophisticated the general public can be when it comes to sports statistics and analysis.

If I were to write a manual on how *not* to conduct an environmental assessment of a mega-project, the process regarding the Northumberland Strait crossing would be my primary case study. Some very basic questions must be asked about environmental assessment: Is it merely an add-on formality with, at best, the ability to inadequately identify mitigation measures for destructive projects? Is it a process with no real way to halt bad ideas or explore meaningful alternatives? Is it an expensive sideshow? Or is it a tool to point us, our children and grandchildren in the best direction possible?

References

CAFSAC, "Biological and Oceanographic Factors Relevant to Consideration of the fixed link Crossing in the Northumberland Strait," 1989. Canadian Atlantic Fisheries Scientific Advisory Committee. CAFSAC Advisory Document 89/9.

Chomsky, N., "Monday Morning Policy Wonks," 1993. *Harper's Magazine*. Volume 286, No. 1714. March, 1993.

Franklin, Ursula, *The Real World of Technology*. CBC Massey Lecture Series. CBC Enterprises.

PEI Fishermen's Association Ltd., "Environmental Effects of a Bridge Across Northumberland Strait," 1990. Submission to EAP. January 29, 1990.

Rice, J.C., C. Morry, T. Sephton, G. Seibert, S. Messieh, B. Hargrave and R.D. Alexander, "A Review of DFO Concerns Regarding Possible Impacts of a Fixed-Link Crossing of Northumberland Strait," 1989. Canadian Atlantic Fisheries Scientific Advisory Committee. CAFSAC Research Document 89/16.

Government Bias and How to Avoid It: A Numerical Assessment Scorecard for the Fixed Link

Tom Kierans

In March 1988, Public Works Canada (PWC) issued a call for proposals for the long awaited fixed crossing between New Brunswick and Prince Edward Island. Unfortunately, PWC restricted proposals to those for a bridge or a road tunnel. A rail tunnel was excluded "for not providing continuous service." This very questionable reason for excluding a rail tunnel from the bidding raises concerns of pro-bridge bias. For a variety of reasons, rail tunnels are much preferred elsewhere for long sea crossings — in particular, they have a good record for public safety. Yet, after reviewing the bids, PWC announced that the link would be a bridge.

The bridge has caused anxiety because of its potentially significant impacts on fisheries as well as other interests. These public concerns caused the federal Environment Minister to appoint a six-person Environmental Assessment Panel (EAP) in 1989 to review PWC's decision. This writer was engaged by Environment Canada as the tunnel advisor to the Panel.

After a year of public hearings and review, the Panel concluded that the bridge project should "not proceed." The Panel also stated that a rail tunnel should have been considered. Despite these conclusions, negotiations were continued with bridge contractors and a contractor is now preparing for construction, pending final government approval. If a bridge is built and serious accidents occur due to weather conditions, concerns will always be raised about PWC's exclusion of a rail tunnel from bidding.

My objectives in this essay are to discuss PWC's rejection of a rail tunnel, and the Panel's rejection of PWC's proposed bridge; and to propose a numerical scorecard to improve environmental assessments and avoid such publicly undesirable impasses.

PWC and PEI's Crossing Options

Over the years, PEI's crossing options have included a causeway, rail and road tunnels, a bridge, combinations of these structures, and improved ferry service. Factors that should be considered in assessing the options include: safety for those travelling and working on the crossing; fisheries protection; user costs for the proposed fixed crossing service; federal concerns to improve travel but also end ferry subsidies; the need for a fast, dependable crossing system by food and other producers; lifestyle disruption for Island residents by any fixed link; loss of jobs by 651 ferry workers; and tourist trade benefits and impacts from a fixed link.

PWC sees winter hazards on a road bridge as "minimal." In fact, the PWC "project rationale" does not mention safety as a factor and implies that a bridge will offer PEI travellers more continuous safe service than a rail tunnel. However, in 1986-87, 17% of scheduled winter ferry trips were delayed by weather. Even less severe weather than that which stops a ferry will make road travel unsafe on a long bridge. A rail tunnel does not need to be shut down because of the weather. European data shows road car travel is 24 *times* more dangerous than travel by rail.

A significant sign of PWC's pro-bridge bias was its questionably-motivated acceptance of road tunnel bids over rail tunnel bids. Major contractors are aware that long subsea road tunnels require very costly ventilation, that modern tunnel boring machine diameters permit their use only in rail tunnels and that bridge bidders had special geotechnical advantages in this case. These facts explain why only one out of seven bids was for a road tunnel and it had to modify PWC's specifications to have any chance of success.

PWC claimed that one tunnel bid showed "consensus that a tunnel is not a viable response to the requirement of the proposal call." This is obviously untrue because rail tunnels are preferred to bridges and road tunnels for most long sea crossings. The PWC claim indicates a possibility that PWC accepted road tunnel bids only because it was a virtual impossibility for a road tunnel to meet PWC specifications. As expected, PWC rejected this bid and selected three road bridges for further consideration.

In accordance with Canada's Environmental Assessment and Review Process (EARP), a six-person Environmental Assessment Panel (EAP) was appointed to review the impacts from the proposed bridge. Early in its review and public hearings the Panel wrote to PWC requesting "an easily comprehensible risk and benefit diagram" of a bridge and tunnel. In summary, PWC's reply was as follows:

1. "There is no reason to assume that the direct benefits of a bridge or tunnel are different."

2. It is found to be impossible to present costs and benefits graphically."

3. Because there was only one tunnel bid "there is a consensus that a tunnel is not a viable response to the requirements of the proposal call."

PWC's reply to the panel's request is unreasonable because:

1. To state there is "no reason to assume that the direct benefits of a bridge or a tunnel are different" defies logic. Fisheries clearly cannot be affected by a tunnel but can be seriously harmed by a bridge.

2. To say "it is impossible" to produce tunnel-bridge cost-benefit graphs is wrong. Such graphs are routine for all project assessments.

3. The claim that lack of tunnel bids shows "consensus that a tunnel is not a viable response to the requirements of the proposal call" is false. Only one tunnel bid was tendered because all tunnel bids were either eliminated or discouraged.

After public hearings, the Environmental Assessment Panel Report stated as follows: "The Panel agrees that there is a need for improved transportation service between Prince Edward Island and New Brunswick. After careful consideration, however, the Panel concludes the risk of harmful effects from the proposed bridge is unacceptable. The Panel recommends therefore that the project not proceed." Moreover, the Panel stated "that the rail tunnel option should have been given consideration comparable to that given other fixed crossing modes."

Environmental Assessments

The goal of most projects that involve public resources and concerns is to provide social and economic benefits. However, the complex choice of options as well as the many factors to consider, such as safety, health, environmental impacts and cost, make selection of the most beneficial option difficult. To protect Canadians, federal regulations require environmental assessment reviews of projects that may have potentially significant impacts or which cause wide public concern.

Experience shows that such reviews can be long and costly. Their adversarial style often divides those whom they are intended to protect into firm proponents and opponents. In such reviews, proponents state project benefits in words that could have several meanings and opponents similarly emphasize negative impacts. Results are poorly understood by a concerned but confused public, to whom it appears that those with the most staying power win, despite project benefits or impacts. Even when reasonable objectivity is maintained, assessment conclusions are often unclear to the general public.

Errors or misjudgment in estimating the significant weight of a factor or assessing an option for a particular project factor can be costly to public safety, health or the welfare of a community. Prince Edward Island's fixed link is a case in point. Therefore, with all projects that can endanger people or that involve public concern regarding air, water or other resources, the result of a project option's assessment should be stated in terms the general public can "measure."

To gain public cooperation, which is essential in environmental assessments, project benefits should be compared directly with impacts in numerical terms the public can readily grasp. This means that assessment results should be on scorecards that show benefits compared to impacts for all options and for every factor considered. This calls for experience and thought, but it can — and eventually it must — be done. Well-designed assessment scorecards can help to resolve many concerns regarding projects that affect public policy.

Numerical Environmental Assessment Design Criteria:

1. Numerical environmental assessment systems should apply to *all* projects that may affect or be affected by natural and human environments and the steps in conducting numerical assessments should be standardized.

2. In a project option's numerical environmental assessment, each factor considered should be "weighted" as to its percent of relative significance to all factors considered. A project option's benefits should be shown numerically in direct relation to its impacts.

3. A project option's numerical environmental assessment results should be stated on widely acceptable scorecards and represent a consensus of objective judgments of experts for each factor involved.

4. In numerical environmental assessments, assessors should be authorized by responsible public authorities. Consultants used in assessments should be recognizably qualified in their field of experience.

5. Public opinion polls may be used if public concerns are identified as a factor and are suitably weighted. Consensus in estimating the unweighed benefits of an option applicable to any factor may be achieved by polling qualified experts in that field of technical expertise.

6. Numerical environmental assessments systems should apply to existing environment problems such as chronic drought or flood or the absence of a project needed to correct environmental problems like waste disposal.

7. Numerical environmental assessments should express a project's current status only. If a project proposal is changed it may be reassessed to reflect proposed changes.

8. An appeal procedure should be developed for numerical environmental assessments. In numerical assessments, appeals should be simplified due to the ease of concentrating on specific factors claimed in the appeal. Factors can be increased to any level of detail needed.

Numerical Environmental Assessment Definitions:

In the numerical environmental assessment system that is proposed here, the words listed below are defined as follows:

- "Option" means one of a number of ways to achieve a project objective.
- "Factor" means conditions that can affect or be affected by an option.

- "Environments" are factors that represent all conditions, circumstances and influences that affect or can be affected by a project option.
- "Assessment" is a numerical statement of a project option's benefits relative to impacts for each and all factors considered in an assessment.
- "Benefit" is a beneficial change caused by or to an option. "Impact" is an adverse change caused by or to an option.

Assessment subjectivity cannot be eliminated but it can be reduced by training experts employed in environmental assessments, who now use words, to state their judgements numerically. Experts must learn to: express the significance or "weight" of each factor involved as a percent of the total significance of all factors considered; and to express their judgements on the value of the benefits of each option in relation to its impacts for each factor considered on a 0 to 10 scale. Thus, total relative benefits for each option and each factor will represent an option's cumulative score on a scale of 0 to 1000. Over 500 means the option has more benefits than impacts. Under 500 means it has more impacts than benefits.

To illustrate the numerical assessment system proposed here, I have assessed options for the 13 kilometre crossing of the Northumberland Strait between New Brunswick and Prince Edward Island. The options assessed are a bridge, road tunnel, rail tunnel and ferry. A proposed assessment scorecard is attached. The procedure used to score option benefits compared to impacts for the four options and for all factors is as follows:

Step 1: All assessment factors (conditions that affect or could be affected by an option) are arranged in groups. Each factor group is allotted Weight Numbers (WN) that express its estimated percent of environmental significance relative to all factors. Factor groups used here are: Safety & Health (30%), Natural Environments (25%), Social (15%), Economic (25%), Technical (5%). Groups are subdivided and WN allotted to subfactors. Factors are unlimited in number but WNs must total 100.

Step 2: For each factor and all options, benefits are compared to impacts on an unweighed benefit points (UBP) 0 to 10 scale. For example, O = no benefits, 10 = no impacts, 5 means benefits = impacts.

Step 3: To obtain an option's weighted benefit points (WBP) for each of the factors considered, multiply its UBPs by that factor's WN.

Step 4: Each option's Environmental Assessment Rating Number (EARN) is the sum of all its WPBs. 1000 means a perfect benefit score. 500 means the option's benefits = its impacts. Over 500 means the option has more benefits than impacts. Below 500 means more impacts than benefits.

The following table is my Numerical Environmental Assessment Scorecard for the Northumberland Strait crossing options.

FACTORS		OPTIONS							
WN in %	Unweighted Benefit Points=UBP (on a 0 to 10 scale) Weighted Benefit Points=WBP WBP=UBPxWN	ROAD BRIDGE		ROAD TUNNEL		RAIL TUNNEL		FERRY	
		UBP	WBP	UBP	WBP	UBP	WBP	UBP	WBP
30	SAFETY & HEALTH								
	During Construction & Maintenance								
2	- weather conditions	3	6	4	8	8	16	7	14
1	- equipment safety	6	6	7	7	8	8	7	7
	During Operation								
11	- weather conditions	2	22	7	77	8	88	6	66
6	- operator-caused accidents	4	24	4	24	9	54	8	48
	Impact to Structure								42
6	- from floating ice	4	24	10	60	10	60	7	21
3	- from ships	3	9	10	30	10	30	7	4
1	- from others & sabotage	1	1	2	2	8	8	4	
25	NATURAL ENVIRONMENT								
	Effect of Structure on Environment								
1	- atmosphere	4	4	8	8	8	8	9	9
2	- freshwater aquifers	5	10	5	10	5	10	9	18
15	- marine biology	2	30	8	120	9	135	9	135
	Effect of Environment on Structure								
7	- high wind, snow, ice, rust	3	21	6	42	9	63	3	21
15	SOCIAL								
	Legal								
5	- constitutional	9	45	9	45	9	45	9	45
1	- security	5	5	5	5	5	5	5	5
4	- public support	6	24	6	24	8	32	5	20
	Cultural								
2	- recreation & tourism	7	14	8	16	9	18	8	16
3	- historic "way of life"	3	9	7	21	8	24	9	27
25	ECONOMIC COSTS & BENEFITS								
5	- construction	3	15	2	10	5	25	9	45
3	- maintenance & operation	2	6	3	9	7	21	3	9
2	- support services	3	6	7	14	8	16	7	14
2	- short-term employment	9	18	9	18	9	18	3	6
4	- long-term employment	6	24	6	24	7	28	8	32
3	- tourism development	7	21	8	24	8	24	7	21
2	- other development	8	16	8	16	8	16	7	14
4	- improved crossing time	7	28	9	36	9	36	6	24
5	TECHNICAL								
2	- structural integrity	6	12	7	14	8	16	9	18
2	- experience with option	4	8	3	6	8	16	9	18
1	- flexibility of option	4	4	8	8	9	9	9	9
100	**Total Benefits=EARN: 1000***		412		678		829		708
	An option's *Environmental Assessment Rating Number=EARN:10000								

SCORECARD PROCEDURE

Step 1 A factor's Weight Number (WN)= its percent significance to all factors
Step 2 Unweighted Benefit Points (UBP) are estimated on a 0 to 10 scale
Step 3 Weighted Benefit Points (WBP)= UBPxWN
Step 4 Each option's total WBPs = its Environmental Assessment Rating Number

Conclusion

Option scores for the test are as follows: a bridge (412); a road tunnel (678); a rail tunnel (829); and a ferry (708). Of course, the scores are based on the writer's judgement for each option and the factors considered in the Environmental Assessment Panel's 1990 Report. Again, in actual assessments, all panel members will be required to express their judgements numerically using procedures outlined here and on the scorecard.

This numerical assessment test represents the views of one individual only, but it demonstrates the following vital facts:

1. It is possible to develop a system that can avoid PWC's exclusion of a reasonable option based on consideration of *one* factor only.

2. It is possible for the public to see the factors considered, the percent weight attached to each factor as well as the assessors' judgement on a 0 to 10 score for each option and all factors considered.

3. It is possible to focus attention on a particular factor judgement for an option and question it without confusing it with other factors.

4. This test shows that the Panel's judgement to reject the bridge and consider a rail tunnel is well-founded.

In conclusion, the continued ferry option ranked quite high in this numerical assessment, and all the options including the rail tunnel have a very high per capita cost for a relatively small population. Thus, consideration should be given to an improved ferry operation as the most beneficial current option. In any case, it appears that there is a clear need for a new and *impartial* numerical assessment by qualified and objective assessors.

The Process

Jim Brown

"No stone was left unturned."

Former PEI Premier Joe Ghiz, testifying before the House of Commons committee studying the fixed crossing, March 11, 1993.

Joe Ghiz, who built a career out of passing off puffed-up hyperbole as reasoned thought, saved his biggest howler for last when he told a House of Commons committee on March 11, 1993 that "no stone was left unturned" during the lengthy environmental review of the bridge from Prince Edward Island to New Brunswick. Surely if there is one thing that everyone — bridge supporters, bridge opponents, federal court judges, and people who don't give a damn one way or the other — can agree on, it's that the public has no idea what lies under the stones on the road to the fixed link. In fact, it has taken a frustrating amount of effort to get the politicians and civil servants running the show to admit that some of the stones even exist.

This is not an essay about bridges, tunnels or ferries. It is about responsibility and trust. This essay does not grow out of, nor does it presuppose on the part of the reader, any particular bias when it comes to mode of travel between Prince Edward Island and New Brunswick. What it does presuppose is a common desire to be treated openly and truthfully by government, and what it documents is a complete failure by those we have empowered to act on our behalf and live up to the trust we have placed in them.

It is "the process," or, more precisely, "the *public* process," that concerns us here: the sequence of events which began in 1985 — when a rookie Island cabinet minister was approached by a Toronto-based consortium with a plan to build a tunnel between PEI and N.B. — and continues today, with events controlled by public officials and subject to the rules and regulations public involvement demands.

We bring certain expectations to a public process that we don't bring to private enterprises. We expect, for example, to hear the truth. We expect publicly funded assessments and consultants' reports to be accurate, fair, and fully disclosed. And chiefly we expect from the politicians and bureaucrats administering a public process a sense of guardianship; we trust them, perhaps naïvely, to look out for our interests ahead of those of developers or special interest groups.

The guardian of the public trust during the fixed link process has been Public Works Canada (PWC). PWC has set the agenda for the crossing project since taking it over in 1986. It has commissioned the studies, negotiated with the developers, held the open houses and issued the press releases. Unfortunately, as well as overseeing the project on behalf of the public, PWC has also been the chief proponent of the bridge. This fundamental flaw has undermined the entire process and everything that follows herein flows from the fact that PWC holds these two irreconcilable sets of goals. In concentrating on its chosen role as champion of the bridge, PWC has abdicated its primary responsibility to the public it serves, so much so that today it can be said without exaggeration that nothing PWC says about the bridge should be believed. Absolutely nothing.

• • •

"If the project had initially been put to me as a bridge, I would have been much less hospitable to it."

(Former federal Environment Minister Tom McMillan, in an interview with the author, September 1992.)

It is somehow fitting that an irony lies at the very inception of public involvement in the fixed crossing project.

The public, through its representatives, was introduced to the idea of a fixed link by a consortium called Omni Systems Group, which approached Tom McMillan in 1985, shortly after he was sworn in as environment minister, with a plan for a single-track rail tunnel to be constructed beneath the Northumberland Strait. The Omni proposal came unsolicited and McMillan says he was dubious at first: "I wasn't an immediate champion." But the more he thought about the idea, the more he became convinced that the massive problems facing Islanders could only be solved by a massive undertaking. "The option was to become a ward of the rest of Canada."

McMillan took up Omni's case. He arranged briefings for cabinet ministers, MPs from all parties, and their staffs, he arranged meetings with the Atlantic Conservative Caucus and the Atlantic Liberal Caucus. Initially, he says, "there was very little support anywhere politically except in my office," but that quickly began to change. Polls were commissioned by Omni that purported to demonstrate that Islanders were overwhelmingly in favour (about 70/30, McMillan says) of an alternative to ferry service. Both federal and provincial politicians began to see that great short-term regional economic benefits could be reaped during the construction of the mega-project. And, perhaps most importantly, bridge builders began to show interest. "The other companies came forward on an unsolicited basis," McMillan says. "Bridge builders approached me and my office. Things moved fairly quickly. Unfortunately, the initial proposal for a tunnel became less and less attractive to the federal government."

Countless theories have been floated as to why Ottawa ruled out, very early in the process, the tunnel option. Some say a rail tunnel didn't fit the Conservative agenda of transportation deregulation and privatization. Some claim Maritime politicians lobbied hard for a bridge because of its greater potential for creating person-years of employment during construction. Some point to extensive lobbying by the bridge consortia and political influence exerted by two large cement companies. McMillan says it was money. "The technology of tunnel building was not such that they could fix a cost, so the minister of finance and his people said 'No way.'" What's important here is not why PWC decided to eliminate a tunnel as a possible option, but rather the fact that for over two years it was maintained that the tunnel option was still on the table. This deception was extremely important to the early success of the crossing project. It provided PWC with an outlet valve whenever the environmental heat was turned up high. It played a significant role in reducing opposition to the crossing by concerned interest groups like fishers. But the biggest benefit to PWC's game plan was the effect the tunnel deception had on the results of the January 18, 1988 plebiscite.

I remember getting a phone call from a PWC official in late November 1989, almost two years after the plebiscite. He wanted to talk about some comments I had made in a radio editorial about the fixed link. This official said he was getting frustrated with all the heat his department had been taking from media types like me. We didn't understand, he said, that PWC wasn't promoting the bridge, it was merely acting on the wishes of Islanders. "After all, 58% percent of Islanders want a bridge."

Despite the fact that Islanders cast their ballots with the clear understanding that a tunnel was being considered, and despite the fact that many Yes voters have subsequently stated that the bridge option is not acceptable to them, PWC has consistently pointed to the results of the plebiscite as an endorsement of their bridge plans. This has been a very effective tactic, and it is a direct result of the two-year tunnel deception. This model for behaviour — the selective release of information and the warping of data — was developed very early on by PWC and has been applied to virtually all aspects of the process under its control.

There is one more important legacy from the early days of the process that lives on today. While the Omni Systems Group tunnel proposal may have been pushed aside by the bridge proponents, one important element of it was retained. It was the Omni group that developed the model of a privately built public facility, a concept that found immediate approval in Tory Ottawa. On paper, it is easy to understand why the politically conservative were attracted to this scheme. A subsidy expected to remain in perpetuity (Marine Atlantic's yearly operating grant) would be replaced by a relatively short-term (35-year) operating grant to the bridge builder. The builder would finance, construct and operate the facility and, at the end of the 35 years, turn it over to government. That's the theory. The reality is that this model has proved to be an unacceptable hybrid, combining the worst features of public and private enterprise — the legislative fast-tracking abilities of the minister of PWC with the confidentiality accorded to private corporations. Under the Omni model, Public Works Canada officials have, literally, two sets of rules available to play under. It is this strange reality, as much as anything, that has been responsible for PWC's schizophrenic administration of the fixed link process.

One final irony should be pointed out before passing on to the next stage of the process. It wasn't long before Tom McMillan joined the other fixed crossing initiators from Omni on the sidelines. McMillan says he knew at the outset that he might end up paying a price for his support of the link. "I was told by several of my political mentors it would be a political tar baby," he says. His mentors were right. In the federal election of November 21, 1988, Tom McMillan was defeated in his bid for re-election in the riding of Hillsborough. One of the reasons for his defeat was the strong showing by anti-fixed link candidate David Weale.

"I failed to appreciate the extent of public and media unease about the process the federal Department of Public Works was following to assess the environmental impact of the fixed link," McMillan wrote shortly afterwards.

• • •

"It seems downright silly and an incredible waste of public funds and people's time to find, at the end of the day, that one of the reasons for which the government rejects the Panel's recommendations is that the Panel did not have before it the detailed information to which the government was privy and which the government had refused to provide the Panel."

(Madam Justice Barbara Reed, in her Federal Court decision of March 19, 1993, referring to Public Works Canada's rejection of the findings of the Environmental Assessment Panel.)

Public Works Canada determined very early on that the key to a smooth process was the successful manipulation of information. The department would decide what reports were released and when, and, just as importantly, what reports weren't released. As a result, there has been a core of secrecy imbedded deep within this process from the very beginning.

In late 1987, PWC announced it would be holding a series of open houses on Prince Edward Island to allow members of the public to review and comment on a set of consultants' reports called the Generic Initial Environmental Evaluation (GIEE). No information from the reports would be released in advance, and only a condensed version would be available to take home for further study. When Jim MacNeill, the editor and publisher of the *Eastern Graphic* and *West Prince Graphic* newspapers, heard about the PWC plan, he didn't like it much. "I think the timing was deliberate," MacNeill says. "They purposely scheduled the meetings (November 20 to December 1) when people would be preoccupied with other things and they gave very little notice and no information in advance."

MacNeill decided to get a full set of the GIEE reports so that his readers, at least, would be prepared for the open houses. He visited the Northumberland Strait Crossing Project offices in Charlottetown where he was informed by project leader Glenn Duncan that he couldn't take the reports because there was only one set available for the whole Island. When pressed further, Duncan managed to find another set for MacNeill.

In examining the GIEE studies, MacNeill made two very interesting discoveries. "When I compared the total report with the condensed version they were handing out, I found things that were totally at odds." He also realized that a report was missing.

Despite Duncan's claim that only one set of reports existed on PEI, MacNeill tracked another one down at the premier's office. Comparing the premier's set with his own, he found that, not one, but two reports had been kept from him: the studies on economic impact and funding.

When MacNeill wrote his news stories about the GIEE reports, including the ones he wasn't supposed to have, Glenn Duncan publicly accused him of theft. When then-Public Works Canada minister Stewart McInnes was asked to explain the discrepancies MacNeill discovered in the reports, he said it was because MacNeill was looking at early draft copies. A check with the consultants who prepared the reports found that no draft copies had been prepared.

MacNeill's work on the GIEE reports led to a series of 46 articles published over a six-week period. These articles earned MacNeill and his newspapers an Honourable Mention as one of five finalists for the Roland Michener Award for Meritorious Public Service in Canadian Journalism. The experience also led MacNeill to come to a conclusion that has proven to be quite useful over the course of the fixed link process: "From then on," he says, "I didn't accept anything Public Works Canada said."

Jim MacNeill may have been the first, but he was by no means the last journalist to encounter frustration in attempting to pry information about the crossing project from Public Works Canada. Freelance writer Lorraine Begley can produce a stack of letters documenting the delays and run-arounds to which she has been subjected by PWC's Atlantic Region in dealing with her requests filed under the Access to Information Act. Investigations by the Information Commissioner of Canada into the handling of Begley's requests have shown PWC officials "overlooked" documents she requested, displayed an "indifferent approach" to her requests, and subjected her to an "excessive delay which was unreasonable and most unfortunate."

CBC Radio news reporter Sandy Smith experienced many of the same frustrations felt by Begley when he attempted, through Access to Information, to get copies of briefing notes prepared for PWC minister Elmer MacKay and Northumberland Strait Crossing Project leader Greg Vaughn. Despite the fact that the Access to Information Act requires government to provide the information requested within 30 days, or to provide notice that an extension is required, Smith received no reply to his request for three months. He only received the information then, he believes, because he complained about his treatment to the Information Commissioner.

And the frustration doesn't necessarily end when the information arrives. Entire two-page briefing notes received by Smith were completely blanked out, and only the page numbers were left to show that a document had existed. Other documents were missing key paragraphs. In some cases, sentences, or even portions of sentences, were removed, leaving the reporter to try to make sense out of statements like this: "[   ] would raise many questions and would be difficult for the Government to explain."

"Whenever there's secrecy it can lead to suspicion," Smith says. "All of this may be above board, all of this may be fine, but the problem is we don't know because they withhold."

They reason information is withheld is clear: The process, as it has been conducted by PWC, simply cannot endure the degree of public scrutiny that full disclosure would permit. An illustration of this can be found in an official document received by Smith that did not come through an Access to Information request. A federal background paper prepared for MPs by the Library of Parliament was forwarded to Smith by an anonymous source. The seven-page paper gives a brief summary of the crossing project, and attempts to explain government's involvement in it. The matter of the bridge subsidy is also mentioned. PWC officials and politicians have maintained throughout the process that the subsidy paid to the bridge developer will be the same as the cost of subsidizing the Marine Atlantic ferry service. As recently as February 1993, Elmer MacKay told the House of Commons that "... it has been the policy of our government that if the project were to proceed it would be an undertaking of the private sector and at a cost to Canadian taxpayers no greater than the overall cost of continuing the existing ferry service."

The background paper, however, tells a slightly different story. "The bridge will derive revenue from two sources," it states. "One is a $42-million-per-year subsidy from the federal government, payable for 35 years, approximating the annual subsidy that would have been paid to the ferry operator and adjusted, *by approximately 50%*, for various factors ..." (italics added). Nowhere in the paper are these "various factors" explained. Nor have they been satisfactorily explained by any public officials.

Even many bridge supporters, Smith says, are openly questioning PWC's handling of the fixed link. "If you take away my opinion, your opinion, and anyone else's opinion and you just take what is here in black and white, and what is not here in black and white, you have to wonder how wise it is to allow a project to go this way in terms of sharing information with the public," he says, adding, "It can at times seem like a Keystone Kops story."

• • •

> *"The purpose of this newsletter is to report on the issues and events impacting on the proposed fixed crossing of the Northumberland Strait and provide a forum for comment. It does not in any way promote the construction of a fixed crossing, or any particular crossing alternative."*
>
> (Editor's Note from Issue #1 of *Strait Facts*, June 1987)

As much time as Public Works Canada has spent keeping information about the fixed link out of the hands of the public, even more time has been spent shaping the information we do receive. A relatively benign, but nonetheless illustrative example of this was the monthly newsletter *Strait Facts*, published from June 1987 to May 1990 by Sparrow Communications of Charlottetown for PWC.

According to Frank Ricketts, the PEI editor of the newsletter, PWC realized that "most people aren't going to read the reports and they're not going to make the effort to come to the meetings," so *Strait Facts* was created "to keep people informed about the process." He says the editorial intent, at all times, "was to be objective. Everything was facts, it didn't get into arguments."

Despite Ricketts' claims, and despite the goals outlined in his very first Editor's Note, the creators of *Strait Facts* found that the elusive goal of objectivity was somewhat easier to promise than to deliver. In the 17 issues of the newsletter published by PWC, not one negative article about the crossing was published. In fact, in my readings, I couldn't find one negative comment. In the first issue, readers were introduced to a regular feature called "Speaker's Corner" which was intended to capture the "lively debates and discussions, sometimes caustic, sometimes humorous" that have been sparked by the fixed link over the years. Judging by this feature, the debate was pretty one-sided. Every single "Speaker's Corner" published in *Strait Facts* was an argument in favour of the crossing. *Strait Facts'* coverage of the Environmental Assessment Panel (EAP) hearings was also interesting when viewed in the context of the editor's stated goals concerning objectivity. Of the 150 separate presentations made to the Panel by individuals and groups, *Strait Facts* published two: the opening remarks at the first public hearing by PWC project manager Jim Feltham, and the closing remarks at the final hearing by the very same Mr. Feltham. *Strait Facts* was also peppered throughout with little journalistic gems like this one: "Based on an informal survey of visitors to the information office and travelling display, 1,611 are in favour of a fixed crossing, 260 are against, and 283 are neutral. Most of those in favour are Islanders who expressed a 3 to 1 preference for a bridge versus a drive-through tunnel."

While Ricketts admits that everything published in *Strait Facts* was vetted first by project officers Jim Feltham and Glenn Duncan, getting an admission from PWC that the same process was followed in the production of the dozens of consultants' reports published by the department is a different matter. But it was.

This is where the official influence on the process becomes insidious and, potentially, dangerous, and where Jim MacNeill's warning to believe nothing becomes particularly useful. It has been stated, most recently by Madam Justice Reed in her decision in the matter of Friends of the Island versus The Minister of Public Works et al., that over $20 million has been spent by Public Works on studies and procedures directly related to the fixed crossing project. These studies and consultants' reports have been presented to the public as detached, scientific research. In reality, there has been as much subjectivity — not to mention coercion — put into the production of these studies as there has been pure science.

More than one consultant has remarked — off the record — that considerable pressure had been exerted by PWC to arrive at certain favourable conclusions in studies concerning the fixed crossing. One has even said that he was instructed to avoid an entire area of concern in the preparation of his report that became part of the Generic Initial Environmental Evaluation package. He was told to avoid the subject, he says, even though it was relevant to his field of study, because its inclusion in his report would reflect negatively on the project.

The problem with reporting on this is that the same stick used by Public Works to beat favourable conclusions out of its consultants is used to keep them quiet about it when journalists show up asking questions. Most consultants in this country succeed by cultivating good relations with government — their future employment depends on it — so they are less than likely to come clean about the degree of arm-twisting that goes on behind the scenes. That doesn't mean it can't be documented, however.

During this process we have seen two very clear, very open examples of the pressures exerted on consultants to arrive at favourable conclusions. When Strait Crossing Inc. (SCI) was instructed by PWC to submit its engineering proposal to a private firm for study and evaluation, the would-be bridge builder sent it to a Vancouver engineering company called Buckland and Taylor. If the proposal was deemed not to be acceptable, that would be the end of the process; if, on the other hand, SCI's plans were approved, Buckland and Taylor would likely be retained throughout the construction phase as engineering overseers. Not surprisingly, the Vancouver firm approved the package. A similar situation occurred when SCI's plans were submitted for review to PWC's ice committee: Turn the plans down and go home; approve them and you've probably got a government paycheque for a few years. The ice committee, like Buckland and Taylor, saw the light.

Of course, there's another dead giveaway that PWC has been exerting direct control over the outcome of its consultants' reports, and it's this: Almost to a document, studies commissioned by PWC have been ringing endorsements of the project; reports prepared without PWC involvement have been critical of either the project or the process.

We've only really had two lengthy public documents released during this process that didn't come from Public Works Canada. One was Judge Reed's detailed ruling, in which she was very critical of PWC's twisting and bending of the rules of the Environmental Assessment and Review Process (EARP). The other was the Environmental Assessment Panel Report itself.

PWC's response to the release of the EAP Report in August 1990 was a case study in denial. It might be instructive, just in case any readers may have forgotten, to reprint the conclusion of the EAP here:

> The Panel agrees that there is a need for an improved transportation service between Prince Edward Island and New Brunswick. After careful consideration, however, the Panel concludes that the risk of harmful effects of the proposed bridge concept is unacceptable. The Panel recommends, therefore, that the project not proceed.

It's a pretty straightforward conclusion, with little room for misinterpretation. Here's how Elmer MacKay responded: "By providing a detailed assessment of the generic bridge concept, the Panel's report sets out specific environmental criteria that could enable the federal government to take the next step in the process." A little over a year later, in the introduction to the Report of the Environmental Committee, MacKay's interesting reading of the EAP Report was expanded upon:

"In August 1990, the FEARO Panel [EAP] published its report. Their conclusions were that the project should proceed only if certain environmental requirements were met ... The federal government responded to the Panel report and accepted its recommendations." Somehow, through the magic of Public Works Canada, an outright rejection was turned into a conditional approval. And the fixed link juggernaut kept rolling along.

Judge Reed, in her ruling, is quite critical of PWC's handling of the Environmental Assessment and Review Process on several counts. First, for striking the Panel too late in the game to fully assess both the bridge and tunnel options and too early in the game to assess the specific proposal selected for construction. Then, for denying requests from the Panel for more detailed information concerning the project. And finally, for rejecting the Panel's findings, largely because it wasn't privy to the detailed information PWC had at its disposal.

She is also unimpressed with PWC's claim that the process involved in producing the Generic Initial Environmental Evaluation — the hurried open houses, missing reports and conflicting data reported on by Jim MacNeill — satisfied its requirements for environmental review. This claim was boldly proposed by PWC despite the fact that the GIEE reports themselves specifically state that further review is required. "In the present case," Judge Reed writes, "one returns again and again to the fact that the particular assessment which is referred to by Public Works as fulfilling the purposes of section 12 (of the Environmental Assessment and Review Process Guidelines Order) contains within itself caveats to the effect that it was not intended to fulfil that role and expressly stipulates that a specific evaluation of the final design must take place."

It will be very interesting to see Elmer MacKay's reading of the Judge Reed decision.

• • •

> *"Strawberry patch philosophy and rails to trails are not going to do it for future generations of Islanders. We will become much more aggressive in our development approach, or we will be left far behind."*
>
> (*Guardian* editorial, August 3, 1992)

There are two more troubling aspects to this process that should be touched on briefly. The first relates, once again, to the diametrically opposed roles taken on by Public Works Canada: the public guardian and the bridge proponent. By taking sides on this issue, PWC has successfully created the impression, among many members of the general public, that opponents of the bridge are troublemakers; a kind of fringe group that, if not anti-government, is certainly anti-progress. I think what this has done, for some Islanders, is shift the question from

"What do I think about the bridge?" to "Who do I want to be associated with?" And I think that's unfortunate.

Finally, there is the role played by the government of Prince Edward Island over the past seven years. There has been, throughout this long process, some security for Islanders in the knowledge that, no matter how badly PWC was botching things, Joe Ghiz was looking out for our interests with his 10 commandments — his list of the minimum requirements for Island approval of the project. This was our safety net, just in case the tightrope broke. It was supremely disappointing, therefore, to discover after Ghiz left office that his publicly proclaimed neutrality on the subject of the link was just for show. The news that Ghiz wrote to Elmer MacKay following the Environmental Assessment Panel's rejection of the crossing and urged the federal minister to get the bridge back on track brought with it the realization that there is no safety net. There never has been.

Tom McMillan, as usual, led with his heart on the fixed link question and ended up taking his lumps. Joe Ghiz, also in character, led with his nose, sniffing the winds for the scent of public opinion, and walking away unscathed from the entire mess. Hopefully, the rest of us will be as lucky.

Popular Protest

Friends of the Island

Betty Howatt

Friends of the Island traces its history back 20 years to a group called the "Brothers and Sisters of Cornelius Howatt." As a member of the Legislative Assembly at the time of Confederation, Cornelius Howatt fought to retain provincial control of internal affairs and also fought for recognition of the family farm as the backbone of the provincial economy. In 1973 the Brothers and Sisters of Cornelius Howatt exposed the provincial government's shift from that priority to one of tourism. That year witnessed the first wholesale marketing of the soul of "the Island" as a tourism gimmick for the Island's centennial year.

Fittingly, the November 22, 1987 initial meeting of the Friends of the Island was held in the kitchen of Cornelius Howatt's former homestead in North Tryon. Many of the same faces, a little more wrinkled than when the Brothers and Sisters had been active fifteen years earlier, were in attendance.

The meeting was to organize opposition to a fixed link and prepare for the anticipated plebiscite on that issue. It was feared that a fixed link between PEI and N.B. would be built without sufficient investigation into the economic, social and environmental impacts such a project would have on the life of the province.

On December 9, 1987, two days after the plebiscite was called, Friends of the Island held its kick-off rally at the Charlottetown Hotel. It generated not only vital seed money but an interest in and support for the Friends of the Island's "Vote No" campaign, leading up to the January 18, 1988 plebescite.

The plebiscite and its results have always been debated — not the numbers, (59.46% for, 40.21% against) but the intent. Many thought a Yes vote meant more investigation of fixed-crossing options, options which included a tunnel. A number of the 59%ers, particularly fishers, who did not want a bridge, voted Yes believing a tunnel was a real option. The 59% Yes vote was interpreted by the proponents as support for a bridge.

Friends of the Island requested a formal environmental review of any fixed-link option. This request was refused repeatedly by the federal Minister of the Environment, at the time pro-bridge booster and Member of Parliament for Hillsborough Riding (Charlottetown area), Tom McMillan. Following his defeat in the November 1988 federal election, an Environmental Assessment Panel (EAP) was appointed.

From the outset the "Friends" faced a formidable opponent, the federal government in the form of Public Works Canada (PWC). Instead of providing information on the four options of bridge, road tunnel, rail tunnel or improved ferry service, PWC became publicly identified as a bridge proponent and was joined by representatives of the organized business community such as the Chambers of Commerce, and the Tourism Industry Association of PEI. Friends of the Island faced opponents with access to both taxpayers' money and private funds to promote their pro-bridge position. By contrast, Friends of the Island's funds came from individuals, worker organizations and fund-raising projects.

Shortage of funds was a road-block to full participation in the EAP process. Friends of the Island wanted to hire consultants to give unbiased opinions on various aspects of the project and applied to PWC for intervenor funding. This was refused on the grounds that the Panel would be able to hire such assistance as its members felt necessary. Despite this setback, Friends of the Island and its individual members presented several quality papers throughout the year of EAP hearings into a generic bridge design.

When the "Report of the Environmental Assessment Panel" was released in August 1990 and it recommended that the bridge not proceed, Friends of the Island was optimistic. But, hard-earned experience with

PWC had taught caution, too. The caution proved to be justified when, three months later, in November, PWC ignored the recommendation and proceeded with plans for a bridge. Friends of the Island undertook legal action to force a review of the specific bridge design selected by PWC.

The Canadian Environmental Defence Fund, an organization which helps groups pursuing national environmental concerns, believed Friends of the Island's cause merited their funds and attention. Through the Canadian Environmental Defence Fund, lawyers specializing in environmental law were enlisted and expert witnesses identified. Friends of the Island's case proceeded to the Federal Court of Canada on March 1, 1993. Much of the money needed for the court action was obtained from the unions of the Marine Atlantic ferry workers through a wage check-off agreement. The decision handed down on March 19, 1993 supported Friends of the Island's assertion that a full environmental assessment is needed before the project can proceed.

Meanwhile, other projects were initiated, such as the post card mail-in campaign and, in the summer of 1992, generous individual and union donations allowed Friends of the Island to open, on a temporary basis, an office to gather information, address public concerns and do media work.

Friends of the Island has brought together a coalition of groups opposed to the bridge project. Among the groups represented are: the Environmental Coalition of PEI, the PEI Fishermen's Association, the Social Action Commission of the Roman Catholic Diocese of Charlottetown, the PEI Federation of Labour, the PEI New Democratic Party, Students of Seattle (from Montague Regional High School), Students for Environmental Action (from Bluefield High School) and the various unions associated with Marine Atlantic ferry workers. Friends of the Island, in consultation with coalition members, acts as spokesperson.

A rally, in which Friends of the Island and its coalition partners participated, was held in November 1992, in Borden. An overflow crowd pressed into the Legion Hall to protest, applaud speakers and cheer entertainers who spoke and sang their resistance to the mega-bridge project.

Labour Organizations

Gerard Sexton

Beginning as far back as the 1950s, labour organizations have opposed a fixed link. Although other strait-crossing projects failed and the idea of a fixed link became connected in peoples' minds with electioneering, when the current proposal surfaced, labour organizations immediately recognized the determination of the proponents to see this one through.

Much of the money to oppose the fixed link originates with Marine Atlantic's largest union, the Canadian Brotherhood of Railway Transport and General Workers (CBRT&GW). Local 127 of the CBRT&GW represents close to 400 employees at the Borden-Cape Tormentine ferry service. Two other unions, the Telecommunications International Union and the Brotherhood of Maintenance Way Employees plus an assortment of non-unionized personnel, make up the 651 workers on this Marine Atlantic ferry service. It is these workers who will be without jobs if the bridge proceeds.

The strain that opposition to the link has placed on the union coffers has been large and long. As principal financier of the anti-link campaign, the Borden local has been forced to fund raise to maintain finances at a reasonable level. Local 127 has sponsored dances at the Borden arena for several years, held lotteries and sells anti-link tee-shirts in an ongoing effort to replenish their constantly called-on coffers. These measures are especially galling considering the funds available to the pro-bridge side from the business community and the federal government.

Promotion of the fixed link is inspired by business, and labour has never been able to credibly take on the business community. For example, the Charlottetown and Summerside Chambers of Commerce were given $49,000 to conduct a study promoting the fixed link, money which came from the federal government through the Atlantic Canada Opportunities Agency (ACOA).

A similar amount of money in the union's hands could have resulted in significantly reduced support for the bridge project. On the unions' side every dollar spent was a scramble and the inequity of this government's largesse has always been a sore point.

As well as being a member of Friends of the Island and active in their anti-bridge projects, CBRT&GW initiated the costly process of launching a court challenge which called for public hearings into the environmental appropriateness of the specific bridge proposal. The decision by the union to approach the Canadian Environmental Defence Fund to take on this case was a difficult one to make. However, members of CBRT&GW felt anti-bridge forces were so severely out-manipulated by both the federal government and pro-bridge advocates that a hearing in the Federal Court was the only solid influence they could have on the outcome of the bridge project.

Friends of the Island, in partnership with their coalition members, launched the court challenge. CBRT&GW Local 127 members agreed to a $100 check-off from their pay cheques over a 10 month period and a reassessment after that time. Recognizing both the importance of and the high costs associated with a court challenge, CBRT&GW's membership pledged that $40,000 would be raised in this way. An appeal to all CBRT&GW locals across the country is expected to bring in another $30,000. The retainer cost for the lawyers was previously paid by Local 127.

Although the Canadian Labour Congress (CLC) has passed a resolution in support of CBRT&GW's struggle and has paid for a study of the economic impacts to Prince Edward Island of the withdrawal of Marine Atlantic, they have made no other monetary contribution. The CBRT&GW has received moral support from other unions within the Prince Edward Island Federation of Labour, and the President of the Federation, Sandy MacKay, has been a strong and active voice.

One of the greatest problems in labour's struggle to defeat the bridge proposal is that of labour solidarity. The most obvious example is the pro-bridge support of the building trades unions and the Canadian Federation of Labour, which formed several years ago after a split with the CLC. This organization has been a vocal advocate of bridge construction.

This breach in labour's tradition of solidarity, as one group of workers battles another to gain employment, has been arguably the most powerful weapon in the pro-bridge propaganda armoury.

The Prince Edward Island Fishermen's Association

Ansel Ferguson

The Prince Edward Island Fishermen's Association (PEIFA) has never taken a position on the issue of a fixed link with the mainland. If a fixed link is to be built, the PEIFA supports the tunnel option. Indeed, in the 1988 plebiscite, many fishers answered "yes" in response to the question which asked if they were in favour of a fixed link. They voted this way because they favour a tunnel as the most appropriate fixed crossing option.

The Fishermen's Association has supported the Friends of the Island in their request for a full environmental assessment of the specific bridge design. They have taken the position that if PWC refuses to request that Environment Canada conduct a full environmental assessment and baseline studies which would determine what species and habitats are present in the strait, then fishers want unlimited compensation unrestricted by area or time. They are also demanding that no contract be signed until this stipulation is in place. With this request, Northumberland Strait fishers are reflecting a realistic fear that if the bridge damages the fishery and no baseline studies are available to prove that certain changes have taken place, then fishers will be denied compensation.

In March 1993, three members of the PEIFA were joined in Ottawa by representatives of the Maritime Fishermen's Union from New Brunswick and Nova Scotia to present this position to the House of Commons Legislative Committee investigating Bill C110.

Both these groups as well as other fishers' organizations in New Brunswick and Nova Scotia and Fishermen's Association members acting individually have expressed their concerns about bridge construction to the 1990 Environmental Assessment Panel (EAP) on several occasions.

An anti-bridge demonstration held in Borden in November 1992 was well attended by individuals from the PEIFA, although the Association did not have a formal presence there. A large number of fishers, acting as individuals, are also members of other anti-bridge organizations such as Islanders for a Free Ferry Service and Friends of the Island.

Environmental Coalition of PEI

Sharon Labchuk

Like many other organizations opposed to the link, the Environmental Coalition of PEI (ECO-PEI) made presentations to the Environmental Assessment Panel (EAP). Early on in the fixed link campaign, it became clear that PEI's largest newspaper had taken a pro-link stance. Those who opposed it had to find their own means of communicating with the public. ECO-PEI's newsletter, *ECO-NEWS*, was distributed widely across the Island, and became a useful tool in balancing the onslaught of propaganda from government and business interests. An in-depth article by Daniel Schulman that appeared in *ECO-NEWS* was later reprinted in *Islandside* magazine, thus receiving even wider distribution.

ECO-PEI's activities have sometimes become difficult to separate from the activities of Friends of the Island. When members of the "Friends" became exhausted and needed a break, ECO-PEI stepped in and carried the load. We wrote a brochure detailing environmental and economic effects of a bridge and distributed it at shopping malls and street corners. This brochure generated extensive media coverage and public debate when PWC disputed their own statements on wind speeds.

Another unexpected source of local, national and international media attention was ECO-PEI's connection to a Japanese environmental group, the People's Alliance to Protect the Japanese Inland Sea Area. Members of the People's Alliance had asked to be kept informed of PEI environmental issues — they had become concerned about these issues when visiting the Island. Their campaign in Japan against the link served to bring the issue to the notice of Canadians outside of the Atlantic area.

Dr. Irené Novaczek, an ECO-PEI member and a representative on the Oceans Caucus of the Canadian Environmental Network, initiated a postcard campaign to oppose the proposed link. Thousands of postcards expressed opposition to a bridge, and were handed out all over the Maritimes. Because Islanders had never been permitted to vote on the bridge proposal, the purpose of the cards was to show politicians just how many people were opposed to it.

Meanwhile, after a short respite, members of Friends of the Island had become active again and hired a coordinator. This action coincided with a critical time for the opposition forces — the government would soon select a company to build a bridge and was anxious to have the contract signed. Friends of the Island needed a visible, downtown Charlottetown office space. A number of landlords were approached and all refused to rent to Friends of the Island. So Friends of the Island was invited to share the ECO-PEI office for as long as necessary.

When it became apparent that the anti-bridge organizations would need legal representation, ECO-PEI helped contact and secure support from the Canadian Environmental Defence Fund. Later, we became one of the supporting organizations for the Friends of the Island court action. Members of ECO-PEI also helped organize the anti-bridge rally in Borden in November 1992. We continue to lobby and some of our members participated in the court action. One, Dr. Irené Novaczek, will be sitting on the Environmental Committee if the bridge project goes ahead. In the meantime, Friends of the Island's funding to operate an office ran out in the fall of 1992. The office of ECO-PEI remains a place where the public can contact the anti-bridge lobby.

Islanders for a Free Ferry Service

Cornelia Howatt

The year-long career of Islanders for a Free Ferry Service has been one of intense frustration. The group came together in February 1992 with the straightforward objective of revealing to all with eyes to see that the federal and provincial governments, in particular Public Works Canada minister Elmer MacKay, were parading about in public in a shocking state of undress. Thus the principal objective of our group has been to present the bare facts — to urge the public towards the realization that the Emperor, in fact, has no clothes.

I mentioned the word frustration. This is an understatement when describing the state of mind, over the past few years, of those of us opposing the construction of the bridge. Our opponents insisted that we be seen a certain way; in essence, we were effectively thrust into a ghetto of perception and denied exit. The promoters of the project, as well as many people in the media, depicted us either as romantic idiots who would ban the automobile and return the Island to the horse and buggy days, or as environmental zealots who would go to any lengths to spare the life of a single fish. Either way, any objections we raised to the bridge, no matter how well-founded, were attributed to our perceived bias and routinely disregarded. The formation of Islanders for a Free Ferry Service represented a bold effort to break out of the ghetto!

Although many of us had opposed the bridge initially because of social or environmental concerns, we became increasingly convinced by the end of 1991 that the whole scheme was a piece of financial and economic foolishness. Stated in its most obvious form, there was no correlation whatsoever between the statement by Elmer MacKay that $40.5 million a year represented a fair assessment of the ferry subsidy, and the claim by Marine Atlantic in its March 1990 submission to the Environmental Assessment Panel that the actual subsidy the previous year had been $21 million. Moreover, Marine Atlantic had gone on to state that it could operate the ferries over the prescribed 35-year period with an annual subsidy in the range of $25.2 million to $28.1 million, maximum, including capital costs such as new ships. "Portions of the studies [carried out by PWC], as they relate to Marine Atlantic, do not reflect reality," declared the exasperated crown corporation.

We hit upon the idea of advocating a *free* ferry service as a means of pointing out the glaring discrepancy between the contending claims of minister MacKay and Marine Atlantic. "Look," we reasoned, "since the annual revenue brought in by the ferry tolls is $12-$13 million, and since this roughly equals the difference between the projected ferry operating costs and what Mr. MacKay is willing to pay, why don't we forget about the bridge and simply request a *free* ferry service?" It was a matter of simple arithmetic! We claimed support for this scheme from none other than former Premier Joe Ghiz, who as Opposition Leader on the Island in 1984 had publicly supported free service as a constitutional right. "Now is the time for the Island government to go after Ottawa to drop all ferry charges," he is reported to have declared in a speech to the West Prince Chamber of Commerce on November 28, 1984.

Islanders for a Free Ferry Service was launched officially at a press conference at the Charlottetown Hotel on February 27, 1992. The group was particularly active during the following three months. Public statements were issued raising further objections regarding the financing of the bridge, pointing out that the tolls could go much higher than officially claimed, asserting that the repair and maintenance costs of the bridge had been gravely underestimated, and stating that no consideration had been given to the potentially astronomical price of decommissioning the bridge at the end of its use. On April 9, 1992, we called for a second plebiscite, this one based on a specific bridge design.

Our group was again very active in the fall and early winter of 1992, especially in the period immediately before and after Premier Ghiz signed, on behalf of all of us, our province's official consent to this dubious deal. We claimed that, as a lame-duck premier who had already announced his intention to resign, he had no right to do this. We also pointed out that despite the ongoing opposition of a good 40% of the Island's people, including almost all fishers, the bridge issue had not once been accorded serious debate in the Island's Legislative Assembly. In recent months (early 1993), Islanders for a Free Ferry Service has taken a back seat to the federal court challenge initiated by Friends of the Island.

What, then, has been accomplished in the past year? I would say two things. First, more people than before can see that we have an Emperor and his cronies going about in a glaringly exposed state. Second, many of us are now convinced that the best option for the Island is a provincially-operated ferry system, similar to that in British Columbia. Should the bridge project fall under the weight of its inherent contradictions, which appears increasingly likely, then we should hasten to embrace Elmer MacKay's extremely generous $40-something-million annual figure! How can he deny to the Island people what he is so anxious to grant to a private company? Given this amount for 35 years, we could operate and improve the ferries and have hundreds of millions left over. Indeed, we could in all likelihood make *free* ferry service a reality ...

In the end, though, our group remains frustrated. Although the federal government has never refuted our claims, Elmer MacKay and others continue to state, over and over, that their constantly escalating figure (at last report more than $42 million a year) represents the true ferry subsidy. And the press continues to accept such statements at face value, without so much as the addition of the telling words "claimed by," used so often in the past to marginalize our own assertions.

At the time of writing the fig leaf remains precariously in place. But for how long?

Contributors

Kevin J. Arsenault was born in Charlottetown, PEI, and raised on a farm in Maple Plains. He received a B.A. in Religious Studies from UPEI and an M.A. in Theology from the University of Windsor, and is currently completing a Ph.D. in Social Ethics at McGill University. He is the Executive Secretary of the National Farmer's Union and Vice-President of Rural Dignity of Canada.

Erik Banke worked on the 1988 MacLaren Plansearch Limited report, "Study of the Sea Ice Climate on the Northumberland Strait," prepared for the Department of Fisheries and Oceans, Dartmouth, N.S. He also authored "Ice pile-up at Stonehaven Harbour, New Brunswick," for the National Research Council of Canada and is currently working with SeaConsults in Vancouver.

Lorraine Begley is a writer and researcher living in Argyle Shore, PEI.

Jim Brown is a journalist and broadcaster who lives in Montague, PEI.

Cooper Institute The Cooper Institute is an organization dedicated to adult education and research. It works with primary producers, workers and other organized groups who are working for social change.

Michael J. Dadswell was born in Saskatchewan, but his first memories are of PEI, where he lived for the first two years of his life. He graduated from Carleton University with a Ph.D., and finally settled in St. Andrews, N.B., where he worked for Huntsman Marine Centre and the Department of Fisheries and Oceans. He now teaches in the Faculty of Biology at Acadia University in Wolfville, N.S., works to help fisheries in the Maritimes and runs a scallop farm in Mahone Bay, N.S.

Donald Deacon was born in Toronto, Ont. and graduated from the University of Toronto in 1940. He served in the R.C.A. during World War II and was awarded the Military Cross. After the war he joined the stockbroking firm of F.H. Deacon & Co., and from 1946-68 was involved in the financing and organization of a variety of businesses across Canada. In 1967 he was elected to the Ontario Legislature, where he served in Opposition for eight years. In 1975, he returned to the Deacon business. He retired in 1980 and eventually moved to Charlottetown, PEI, where he established Atlantic Ventures Trust to invest in emerging business in Atlantic Canada. He served as Chairman of the Atlantic Provinces Business Council from 1987-89 and was a member of the Advisory Board of ACOA. He was appointed a Member of the Order of Canada in 1987.

Martin Dorrell was born in Montreal in 1947. Educated in Ottawa, he began his career in journalism at *The Globe and Mail* in 1970. He worked there as a general reporter, feature writer and copy editor before moving to the Island in 1974. For ten years he worked at CBC Charlottetown as a radio producer and reporter. Since 1986, he has taught journalism at Holland College in Charlottetown.

Ansel Ferguson has fished out of Victoria Harbour, on the north shore of the Northumberland Strait, for the past 32 years. A past President and longtime board member of the PEI Fishermen's Association, he is currently a member of Friends of the Island and Islanders for a Free Ferry Service.

Sharon Fraser is a writer and broadcaster living in Halifax. In the mid-'80s, she lived in PEI, where she was editor of the newspaper *Atlantic Fisherman*, which covers the commercial fishing industry. Following that, she was editor of *Atlantic Insight*. She has taught courses in journalism both at Mount St. Vincent University and in the journalism program at the University of King's College. Her freelance work is published regularly in *This Magazine, The Canadian Forum, The Atlantic Co-operator*, and *Catholic New Times*. She has also done commentary for *The Globe and Mail, The Daily News* (Halifax) and the CBC.

Owen Hertzman is as Assistant Professor in the Atmospheric Science Program in the Department of Oceanography at Dalhousie University in Halifax. He completed his B.Sc. in Engineering Physics and M.Sc. in Climatology at UBC and, after working as an agricultural meteorologist, he completed his Ph.D. in Atmospheric Science at the University of Washington in Seattle. His area of specialization is marine meteorology, particularly winter storms and the rain and snow regions they contain. He has contributed to recent books on the regional climate of Atlantic Canada, and has served for the past two years on the Nova Scotia Department of the Environment's Clean Air Task Force.

Betty Howatt is a founding member of Friends of the Island, and served as Chair and President of that organization from November 1988 to April 1992.

Cornelia Howatt is yet another incarnation of the spirit of Cornelius Howatt, the 19th-century anti-Confederate Islander who held true to his principles to the bitter end. His example inspired the Brothers and Sisters of Cornelius Howatt, a small but hardy band of his compatriots who used his name a century later (in 1973) to rally Islanders to think and act independently.

Tom Kierans, P.Eng is a native of Montreal and a McGill graduate in Mining Engineering. He has prospected for minerals across Canada and been employed as a mine supervisor, safety engineer and mine designer. He had senior responsibility for mining, safety and environmental protection during the Churchill Falls project construction, and was then invited to be an engineering professor at Memorial University of Newfoundland and an editor for the American Society of Civil Engineers' manual on the design of nuclear energy facilities. Tom Kierans was also the first Director of the Alexander Graham Bell Institute at the University College of Cape Breton. He lives in St. John's, Nfld., where he is a consultant in mining, water supply and environmental protection. He also founded and manages a company that designs floating structures, and another company engaged in land development in the St. John's area. He is a member of an environmental protection board dedicated to cleaning St. John's harbour.

Sharon Labchuk is co-ordinator of the Environmental Coalition of PEI, a group active on a variety of environmental issues.

Irené Novaczek was born Irene Hall in Masselburgh, Scotland and emigrated as a child to PEI in 1958. After graduating from King's College, Halifax, she spent a year at UBC, then travelled to New Zealand as a Commonwealth Scholar to do a Ph.D. in Marine Ecology. She eventually returned to PEI to work as a research associate at UPEI. She has published widely on many topics, including seaweed ecology, biogeography, acid rain, toxic plankton, shellfish physiology, shellfish processing and small business development. She is currently Co-chair of the PEI

Environmental Network and Chair of the Ocean Caucus of the Canadian Environmental Network. Much of her time is spent with community groups, helping people to understand and critique environmental impact reports and scientific studies.

Joseph H. O'Grady was born in Charlottetown, where he attended city schools and graduated from UPEI with a degree in Business Administration. After working for a time with the federal government, he left the Island to attend graduate schools in New England. He earned Masters degrees in Business Administration and Economics. Since 1983 he has lived in Burlington, Vermont, where he is a professor of Business Administration. He maintains close ties with the Island.

Reg Phelan is a farmer and carpenter from Byrne Road, PEI, and a member and past President of Local 103 of the National Farmers Union. He is currently studying social movements and the land on PEI for his M.A. thesis at Saint Mary's University in Halifax.

Daniel Schulman holds a B.Sc. in Environmental Sciences from the University of East Anglia (England) and an M.Sc. in Soil Science from the University of Guelph (Canada). He has 10 years of experience in environmental science, environmental technology and environmental activism. He currently teaches the Environmental Technology Program at Holland College and lives in Bonshaw, PEI with Pat and Ben.

Gerard Sexton is a National Executive Board Member of the Canadian Brotherhood of Railway Transport and General Workers. For the past 11 years, he has been Local Chair of Local 127 (Borden). For the past 10 years, he has also served as an Executive Board Member of the PEI Federation of Labour. He is a member of Friends of the Island.

Peter G.C. Townley received B.Sc., B.A. and M.A. degrees from the University of Guelph, and a Ph.D in Economics from Queen's University. He has been an economist in the Department of Economics at Acadia University since 1981 and has published articles on public and private pension plans, annuity markets and cost-benefit analysis. *Economic Policy Analysis*, a textbook he co-wrote, will be released this summer, and he is currently writing a textbook on cost-benefit analysis in a Canadian context with Wade Locke of Memorial University.

Select Subject and Name Index

THE BEST OF RAGWEED PRESS

The Apple A Day Cookbook, **Janet Reeves** This elegant but practical cookbook has something for everyone. From appetizer to aperitif, salad to soup, beverage, bread and breakfast to the main course meal, an "apple a day" will never be the same! **ISBN 0-921556-32-2 $13.95**

Dancing at the Club Holocaust: Stories New & Selected, **J.J. Steinfeld** This is Steinfeld's most powerful collection, gathering the best of his writing — stories about North American Jewish experience. J.J.'s gift of dark humour and psychological insight have never been so illuminating and impressive.
ISBN 0-921556-30-6 $14.95

Lnu and Indians We're Called, **Rita Joe** Micmac poet Rita Joe's third collection of poetry follows and expands upon her desire to communicate gently with her own people and to reach out to the wider community. She invites us to "Listen to me, the spiritual Indian." **ISBN 0-921556-22-5 $9.95**

My Broken Hero, **Michael Hennessey** These stories takes us from the precarious innocence of the 1940s to the tumultuous present day on Prince Edward Island. Hennessey breathes life into his "broken heroes" and treats us to a rare vision of the Island. **ISBN 0-921556-24-1 $10.95**

Nine Lives: The Autobiography of Erica Rutherford, **Erica Rutherford** For fifty years, artist Erica Rutherford, who was born a male, struggled with the conviction that she was the wrong gender. At fifty, she finally made the decision to live her life as a woman. Now, at the age of seventy, she tells her life story, eloquently affirming the variety and vitality of human possibility.
ISBN 0-921556-36-5 $12.95

The Storyteller at Fault, **Dan Yashinsky** A masterful tale of adventure, wit and suspense, by an accomplished raconteur. Folk literature and oral traditions from around the world are woven into a colourful tapestry that is a whole new tale in itself. **ISBN 0-921556-29-2 $9.95**

To Scatter Stones, **M.T. Dohaney** With humour and insight, the legacy of *The Corrigan Women* (Ragweed, 1988) is carried into the present day. Tess, the last surviving member of the Corrigan clan, returns home to the politically Conservative Cove to run for the Liberals as the first-ever "petticoat candidate" in the upcoming provincial election. **ISBN 0-921556-23-3 $10.95**

Toward a New Maritimes: A Selection from Ten Years of New Maritimes, **Ian McKay and Scott Milsom, Editors** The writing gathered here offers an accessible, unorthodox look at the historical, economic, cultural, and personal forces at work in the Maritimes. Includes award-winning investigative journalism.
ISBN 0-921556-34-9 $18.95

RAGWEED PRESS books can be found in quality bookstores, or individual orders may be sent prepaid to: RAGWEED PRESS, P.O. Box 2023, Charlottetown, Prince Edward Island, Canada, C1A 7N7. Please add postage and handling ($2.00 for the first book and 75 cents for each additional book) to your order. Canadian residents add 7% GST to the total amount. GST registration number R104383120.